Adaptation in Metapopulations

Adaptation in Metapopulations

How Interaction Changes Evolution

MICHAEL J. WADE

The University of Chicago Press | Chicago and London

Michael J. Wade is distinguished professor of biology at Indiana University, Bloomington. He is coauthor of *Mating Systems and Strategies*.

The University of Chicago Press, Chicago 60637
The University of Chicago Press, Ltd., London
© 2016 by The University of Chicago
All rights reserved. Published 2016.
Printed in the United States of America

25 24 23 22 21 20 19 18 17 16   1 2 3 4 5

ISBN-13: 978-0-226-12956-3 (cloth)
ISBN-13: 978-0-226-12973-0 (paper)
ISBN-13: 978-0-226-12987-7 (e-book)
DOI: 10.7208/chicago/9780226129877.001.0001

Library of Congress Cataloging-in-Publication Data
Wade, Michael John, 1949– author.
    Adaptation in metapopulations : how interaction changes evolution / Michael J. Wade.
        pages cm
    Includes bibliographical references and index.
    ISBN 978-0-226-12956-3 (cloth : alkaline paper)—ISBN 978-0-226-12973-0 (paperback : alkaline paper)—ISBN 978-0-226-12987-7 (e-book) 1. Adaptation (Biology) 2. Population biology. 3. Ecology. I. Title.
    QH546.W23 2016
    578.4–dc23

                                                                              2015031811

♾ This paper meets the requirements of ANSI/NISO Z39.48-1992 (Permanence of Paper).

To David Evan McCauley,
my career-long friend and collaborator

Contents

1   Introduction

The central question guiding my research throughout my career has been this: *How is the process of adaptation different if the members of a population live clustered in small groups instead of being homogenously distributed like grass on a lawn?* The field is called "evolution in subdivided populations" or "adaptation in metapopulations." It has led me to investigate a diverse array of topics, including group selection, family selection, kin selection, and sexual selection, as well as speciation genetics, maternal and paternal genetic effects, and host-symbiont coevolution. In my lab, my students and I have approached these topics using a combination of theoretical, field, and laboratory studies and a diversity of living systems ranging from our laboratory model of flour beetles in the genus *Tribolium* to other animals, plants, and microbes. Through the generosity of the National Science Foundation Opportunities for Promoting Understanding through Synthesis (OPUS) Program, the Sabbatical Scholars Program at the National Evolutionary Synthesis Center, and the Sabbatical Leave Program at Indiana University, I have had the opportunity to write a conceptual, historical synthesis of the findings from these studies and the relationship between adaptation in metapopulations and broader questions in evolutionary genetics.

This first chapter is an overview of and introduction to

concepts considered in greater depth in later chapters. I first provide a bit of personal background about my family and my schooling, since both influenced my career and shaped my research interests.

The organization of later chapters follows the flow from questions and concepts to laboratory experiments and to field studies. The majority of my mathematical models developed from the physical activity of moving flour beetles within and among families (kin selection) or within and among populations (group selection) and the synthetic activity of analyzing data. The dates on some theoretical publications precede those of the experimental works that inspired them, but only because the design, execution, and analysis of experiments is much slower (and generally more tedious) than working out the results of mathematical models.

New data are included in later chapters; data from experiments that went unpublished, owing to interruptions from teaching, administrative tasks, or lapses in funding, as well as to the dispersion of young collaborators away from my lab to career opportunities elsewhere. Published or not, results from these studies became part of lab lore and influenced our thinking and the planning of subsequent research.

Personal Background

In 1949, I was born in Evanston, Illinois, the oldest of eight children, into a fundamentalist, Irish Catholic family. I spent most of my waking hours in the swamps near our home in Westport, Connecticut, collecting frogs, toads, turtles, snakes, and butterflies at Lee's Pond, Willow Brook Cemetery, and the banks of the Saugatuck River. My parents encouraged our interest in nature by allowing us to keep anything we could catch and my dad built elaborate circus wagons, wheeled cages, for moving our menagerie in and out of the garage. My childhood ambition was to become not a scientist, but rather curator of reptiles at the Bronx Zoo, inspired by Marlin Perkins and his show, *The Wild Kingdom.*

Although neither of my parents had completed college, my mother emphasized our education above all else. (Long after we left home, my mom returned to college and completed her degree at St. Joseph's University of Pennsylvania in 1983 at the age of 61.) And, although she did not believe in evolution herself, she introduced us to it by reading us "dinosaur books" at bedtime. Each summer, we took family trips to Yale's Peabody Museum, where we spent our allowances on rubber dinosaurs and the army men to fight them.

In 1963, my father was appointed director of research for the Triangle Broadcast Center and we moved from Connecticut to Drexel Hill, Pennsylvania, a Catholic enclave near Philadelphia. I attended St. Joseph's Preparatory High School and my endless hours in the swamps were replaced by 5–6 hours of homework each night; a test every week in every subject; and daily quizzes in Latin and math. The Prep faculty had two of my best teachers, Mr. Earl Hart in honors mathematics and Stephen A. Garber, S. J., in honors chemistry. In 1967, I entered another Jesuit institution, Boston College, on scholarship. College was easy compared to high school and a year of advanced placement credit allowed me the freedom to experiment with majors in chemistry, English, sociology, mathematics, and biology. I spent my senior year writing plays, oil painting, and learning Boolean algebra as a "Scholar of the College" before graduating in 1971 with a double major in mathematics and biology. Somewhat unfocused, I applied to law school as well as to graduate schools in anthropology, linguistics, and evolution and ecology. With guidance from my biophysics professor, Dr. Donald J. Plocke, S. J., I applied to and was accepted by the Department of Theoretical Biology at the University of Chicago, where I was supported by a four-year National Institutes of Health (NIH) Graduate Training Fellowship (1971–1975). It was amazing to get paid to go to school, especially to a doctoral program uniquely suited to my combined interests in biology and math. The Chicago faculty applied mathematical models to everything from development to neuroscience to biological clocks. My background in natural history caused me to gravitate toward ecology and evolution. My dissertation was coadvised by Drs. Thomas Park, founder of the field of laboratory ecology and soon to be professor emeritus, and Montgomery Slatkin, a biomathematician and beginning assistant professor. Both were challenging and supportive mentors who guided my earliest ventures into group selection and the evolutionary genetics of metapopulations (chapters 3 and 4).

Interactions and Context

In graduate school, I learned for the first time that Darwinian evolution required variation, replication, and heredity and that any system whose units had those properties could evolve (Lewontin 1970b; later, Maynard Smith 1976) and that selection could operate simultaneously at more than one level and in more than one direction. This made it particularly difficult to determine a priori whether selection at one level was more efficient in producing evolutionary change than selection acting at another. Quantifying the relative efficacy of multilevel selection required an understanding of selection and heredity at each level. The goal of my research was to develop methods for this kind of multilevel quantification and to test the relative efficacy of the levels of selection.

In my graduate classes, single genes were the primary conveyors of heredity and evolutionary change equaled change in gene frequency. Interactions between genes or between genes and the environment were largely ignored. Individuals could be broken up into their component parts, genes and the environment. Reciprocally, these parts could be summed back together to re-create the individual. In a world without interactions, this one-gene-at-a-time evolution made a great deal of sense. But it also implied that the fitness differences at lower levels in the biological hierarchy determined those at the higher levels, because higher level units were merely aggregates of lower level entities. This view that groups were nothing more than the sum of their parts made the levels-of-selection problem a fairly trivial one and it eliminated group selection as a viable adaptive process. It also favored reductionism as the primary research methodology in evolution. Understanding the essence of complex living systems was equivalent to understanding processes at the lowest level of the biological hierarchy; the more you knew about the lowest level, the more you knew about everything else. Of course, my instructors in genetics, biochemistry, and biophysics believed the same thing, with particle physicists winning this race to the bottom of the hierarchy.

My instructors and textbooks asserted that Darwinian selection among individuals, and not among cells, groups, populations, species, or communities, was the strongest and most efficient kind of selection. Natural selection produced adaptations, which were responsible for the fit between an organism's morphology, physiology, and behavior and its environment. Natural selection was privileged in the constellation of evolutionary forces (which included mutation, migration, and random genetic drift) because of its crucial role in adaptation. Kimura's Neutral Theory of Evolution (1968) and the famous paper "Non-Darwinian Evolution," by King and Jukes (1969), both challenged that primacy. These papers were discussed, dissected and found wanting over and over again in lectures and in graduate seminars during my doctoral training. Absent critical evidence, it was relatively easy to tell an adaptive story about almost anything. With data, however, it was much harder to detect the expected signal of natural selection through the noise of mutation and random genetic drift.

War and Poker

To illustrate adaptive evolution as we were taught it, consider the card games war and poker. The received view was that Nature and the evolutionary processes that produced her are much more like the game of war than they are like poker. In war, high cards win and low cards lose, just as genes good for fitness increase in frequency by natural selection and those bad for fitness decrease. In war, good hands have high cards and poor hands have low cards. The quality of a hand equals the sum of its cards. To see the parallel with evolution, pretend that a hand is an individual. The fitness of an individual equals the sum of the effects of its genes and the environments it has experienced. The adaptive quality of an individual can be assessed from the sum of the fitness effects of its genes, just as a hand in war can be assessed from the sum of its high cards. The fittest individuals, favored by natural selection, are those with the best genes; the winner in war is the person with the best hand.

In contrast, if Nature were more like a game of poker, then individuals would not be the sum of their parts. A card in a hand of poker can be labeled good or bad only in the context of the other cards in the hand, the people playing those hands, and often the order in which the hands are dealt and played. An ace is a good card in some hands but it can be a bad card in others. In fact, an ace of one suit can be more valuable than an ace of another suit in the context of a flush. The same hand might win in one game but lose in another; win when held by one player but lose when played by another; or win early in a series of games, but lose later. Context is irrelevant in war, but it is the essence of poker.

A theory of gambling based on unchanging card values would work very well in describing war but poorly or not at all in describing poker. Poker games have emergent properties because interactions between cards, between hands, and between players confer importance to contextual variations in the value of a card. In war, unlike poker, there are no emergent properties. War lacks interactions; it is essentially a context-free game. Where war is mind-numbingly boring, poker can be endlessly fascinating, almost as fascinating as Nature.

Because higher levels of biological organization have emergent properties that are lacking at lower levels, Nature is more similar to poker than it is to war. The genes causing complex human genetic diseases are harder to find and map because genes behave more like the cards in poker than in war. A gene can be good in one background or environmental context, but harmful in another, so the same gene may appear both in healthy and in sick individuals. If upper levels in the biological hierarchy are not simple aggregates of the lower level entities, then selection at a higher level can be stronger than or more efficacious than selection at a lower level. As Feldman et al. (1983, p. 1009) concluded from their review of models of fertility selection based on interactions between mating males and females, "the simplest interactions between individuals in the process of selection can produce evolutionary conclusions not expected from individual fitness models." Unfortunately, we were taught to analyze Nature and adaptation using a genetical theory of war without interactions, not a theory of poker.

Multilevel Selection Theory in the 1970s

Multilevel selection theory was in its infancy when I began graduate school. The year before I started at Chicago, Dr. R. Levins (1970) defined a metapopulation as a *population of populations*. He emphasized that the processes of extinction and colonization of local populations (technically called *demes*) within a metapopulation were analogous to the births and deaths of individuals within a population. In his articles and graduate lectures, Levins emphasized that natural selection occurred within as well as among populations and that a theoretical and experimental framework capable of handling both kinds of selection at the same time was needed. It was the absence of a multilevel selection theory that caught my attention as a graduate student.

Also in 1970, Levins' colleague Dr. R. C. Lewontin (1970b) published an influential review entitled "The Units of Selection." In class, Lewontin lectured about levels of selection below the individual, particularly gametic selection (also known as meiotic drive), where one allele is favored over another when heterozygous individuals form sperm or eggs. A particularly interesting case occurred when males produced more Y-bearing sperm than X-bearing sperm: they produced primarily sons. Such a population could run out of females and go extinct. Lewontin stressed that group selection could be a means of limiting such gametic selection because populations with genetic tendencies to overproduce either kind of sperm would go extinct, while populations without such a tendency would persist. The importance of this example was unmistakable because it appeared on our written doctoral preliminary exam as a problem. Other than the genetics of gametic selection or segregation distortion, however, multilevel selection was rarely discussed in class.

The first class to allow me to develop a research project was Park's Field Ecology course, which was based in the vernal ponds of Chicago's Tinley Creek Forest Preserve. The array of ponds there seemed a natural example of a metapopulation. But my research did not become focused on evolution in metapopulations until my second year of graduate school, when I made the transition from field to laboratory work, using *Tribolium*, Park's flour beetle, as a model system—a transition I discuss in chapter 3.

Interactions and Metapopulations

The standard view of evolutionary theory misses the important fact that genetic adaptation to the internal environment of other genes occurs more rapidly than genetic adaptation to the external environment (Drown and Wade 2014). The "fit" between organisms and the external environment, which motivates the standard theory, is visually striking and can be seen by simply taking a walk in the woods. However, the fit between organism and external environment is no more striking than the complexity of the regulatory "fit" between genes that control development from fertilized egg to adult. ENCODE (Encyclopedia of DNA Elements) has found that, although the human genome consists of a mere 24,000 genes, it has 4,000,000 regulatory regions, which determine where and when each gene is expressed. This is more than 150 regulatory elements per gene. Unlike the external environment, the internal environment of other genes can coevolve, so that a good gene fit on a particular genetic background drags that internal environment along with it to the next generation. In this way, not only do the best genes become more common, but they also make the genetic environments in which they excel more common (Drown and Wade 2014).

Multilevel selection theory offers a more comprehensive account of the adaptive process for complex traits than does standard theory for species whose members are distributed spatially as many, more or less isolated, small populations. It allows us to understand macroevolutionary processes, like the origins of biodiversity and species richness, as the outcomes of microevolutionary processes involving interactions between genes, between genes and environments, between genes in different individuals (like mothers and offspring) and in different species. It is my conviction that the inclusion of interactions of all sorts into evolutionary theory is essential for uniting micro- and macroevolution in an understanding of Nature's enormous diversity of species.

The existence of interactions expands the domain of multilevel selection and, at the same time, constrains individual selection. To return to the card game analogy, the existence of interactions in poker expands the range of strategies and constrains the value of the "high cards only" strategy learned in war. Just as artificial selection helped Darwin think about natural selection acting on individuals, I find artificial group selection useful for thinking about multilevel selection. Consider a trait that affects both individuals and the groups to which they belong, such as the relationship between leaf area and seed yield in plants. At the individual level, larger plants with greater leaf area produce more seeds than smaller plants. At the group level, however, a large plant grows at the expense of its neighbors, shading them with its large leaves, stealing their water and nutrients with its large root system, thereby reducing their seed yield. The best individuals harm their neighbors, reducing average yield of the group. It is this kind of trade-off between an individual and its social partners that decades ago led plant breeders interested in maximizing yield to abandon simple individual selection in favor of group or stand selection. This same trade-off is currently leading animal breeders to adopt group selection as a means of maximizing yield while improving animal welfare at the same time, by breeding kinder, gentler chickens, pigs, and cows (see below). These examples from agriculture and animal breeding are representative of competition between individuals for scarce resources in Nature. This is the kind of competition that lies at the heart of Darwin's evolutionary logic, yet, in experimental systems, the genes affecting group productivity do not respond well or at all to individual selection.

A second example of a trait good for the individual but bad for its group comes from poultry farming to maximize egg yield in laying hens. Economic efficiency dictates that groups of laying hens be housed together rather than individually for ease of handling. But, in a group, a few dominant hens grow larger, obtain more food, and lay more eggs than the hens they subordinate and injure by pecking. Just like the high-yielding plants, the best hens achieve their success by harming others in the group and taking away their resources. In the process of competing, dominant hens lower the egg output of the group as a whole. The previously favored mechanical solution to the pecking problem (considered one of the largest economic problems in the industry) has been "beak trimming," a euphemism for burning the beaks off hens. This procedure physically removes the part of a hen's phenotype that mediates pecking. However, this practice involves the increased labor of cauterizing hen beaks one by one and raises issues with respect to the ethical treatment of animals.

A very different solution to the pecking problem occurred to my colleague at Purdue University, Dr. Bill Muir. Muir used artificial group selection to raise the numbers of eggs laid by a group of hens (group selection) instead of the standard industry procedure of selecting the best individual hens (individual selection). By shifting the focus of selection from the individual to the group, Muir (1996) increased egg yield by 65% in only a handful of generations of artificial group selection, a process used with increasing frequency by breeders since then (Wade et al. 2010). The contrast with the standard individual selection methods at the time was astonishing: industry egg yields were increasing only by 0.5% to 1.2% per year, despite the assistance of the most sophisticated breeding models. Muir's group-selection response was the equivalent of 80–130 generations of artificial individual selection!

Muir was successful because group selection was able to access the genetic variation governing social interactions between hens, heritable variation that was present but beyond the reach of individual selection. By selectively breeding at the group level, Muir created gentler laying hens that not only laid more eggs but also no longer needed beak trimming. Under the industry's long-standing individual selection regime, the best laying hens succeeded at the expense of their cage mates; after group selection, the interests of the individual and the group were aligned. Notably, Muir's large increase in egg laying occurred in poultry strains already subjected to more than 40 generations of elite, pedigree-assisted schemes of individual selection. Clearly, Muir's methods had found an enormous, untapped reservoir of genetic variation affecting egg laying—a reservoir of variation unavailable to individual selection. Muir has said that he was inspired to try group selection in hens by studies in my lab using group selection to increase offspring numbers and decrease cannibalism in flour beetles.

Why did this genetic variation for increased egg laying sit untapped in laying hens (or, for that matter, in our flour beetles)? As I will show later, breeding practices that focus exclusively on the individual and its characteristics allow genetic variation for competitive social interactions with harmful effects on the group to accumulate. Without a mechanism, such as group selection, for screening out group-harmful variation, mutation fills the genetic reservoir of "social" genetic variation just as it fills the reservoir of individual genetic variation when selection is relaxed.

Multilevel Perspective on Heritability

Interactions between genes in the same genome (classic epistasis), between genes in different genomes of the same species ("social" variation based on *intraspecific* indirect genetic effects or IGEs), and between genes in the genomes of different species (*interspecific* IGEs) greatly complicate the concept of *heritability* at *all* levels of selection. Whenever interactions occur within metapopulations, the rate and the direction of selection can vary from one local population to another, not because the external environment varies (although it might), but because the underlying relationship between genes and between genes and traits varies. Whereas individual selection is a variation-consuming force in standard theory, it can be a variation-generating force across a metapopulation when there are interactions, as I will explain in later chapters. Understanding this concept is important to understanding the origins of biodiversity. Some of the empirical and theoretical results from my lab have been surprising only because standard theory has been so narrowly focused on single-gene, Darwinian evolution by individual selection without interactions. Throughout the book, I will use examples from our work to illustrate how group heritability or population heritability (which I refer to as g^2) is different from individual heritability (classically represented as h^2). Because group and individual selection access different pools of heritable variation, we need different symbols to refer to the heritable variation upon which each depends and we need to understand how they affect one another, particularly in the presence of interactions.

Individual heritability (h^2) measures the degree to which offspring resemble their parents. Imagine that one selected the largest individuals from a population where individuals differed in size, and exclusively bred these large individuals so that their offspring formed the next generation. Some of the selected individuals will be large because of their genes, while others will be large because of the favorable environments they experienced when young. Only the *genetic* tendency toward larger size is heritable and transmissible to the offspring; the environmental component is not. As a result, the average size of the offspring ($X_{offspring}$) has to be smaller than the average size of their selected parents ($X_{parents}$), because not all of the parents are large for genetic reasons transmissible to the offspring. Heritability, h^2, measures the fraction of genetic resemblance between parents and offspring. Breeders' experience has shown that, if estimated once, h^2 can be used to predict the response to artificial individual selection for the next several generations, and sometimes for up to 20 generations or more. Breeders have also shown that after long periods of imposing individual selection, h^2 for many traits declines toward zero, and selective gains diminish or come to a halt. Continued individual selection is not as effective once it has consumed the genetic variation in the ancestral population before selection.

Group heritability, g^2, has a parallel meaning for group selection. Imagine a metapopulation from which one selected the largest local populations, and exclusively used individuals from those to establish new populations in the next generation. Some selected populations would be large for genetic reasons, while others would be large because of the favorable local environment where they developed. Because only the genetic tendency toward larger size is heritable, while the environmental component is not, the average size of the offspring populations in the metapopulation next generation should be smaller than the average size of the selected parent populations. Population or group heritability, g^2, measures this fraction of genetic resemblance. Clearly, h^2 and g^2 are similar concepts. However, under some circumstances, g^2 can exceed h^2 (Wolf et al. 1998).

There are additional differences between g^2 and h^2. Our experiments have shown that the properties of g^2 are much more dynamic than those of h^2, which tends to remain more or less constant over time. In newly subdivided metapopulations, the initial value of g^2 tends to be small, but it increases rapidly, within several generations, to fairly high values in response to random drift (Wade 1978a; Wade and McCauley 1980). Additionally, we have found that, for traits like fitness, h^2 can be small, while g^2 is very large. Not surprisingly, group selection is a much more effective evolutionary force than individual selection when g^2 is large but h^2 is small.

Without interactions, per standard theory, g^2 would be only a small fraction of h^2. This almost automatically would make individual selection a much more powerful evolutionary force than group selection. In Nature, however, genetic and environmental interactions are ubiquitous, so that "Context and interaction are of the essence" (Lewontin 1974). With such interactions, g^2 not only can exceed h^2, it can exceed the maximum possible value of h^2. For these reasons, I reframe the concept of heritability as a multilevel phenomenon in this book.

Understanding selection and heritability as multilevel processes fundamentally changes our understanding of variation, replication, and heredity at the individual level. Under the standard view, adaptive evolution occurs along a one-gene, single dimension that "requires first a synthetic treatment of the genotype and then an abstraction of the single system of interest from the complex mass. We cannot reverse the process, in general, build-ing a theory of a complex system by the addition or aggregation of simple ones" (Lewontin 1974, p. 281). If Nature were not complex, she could be dis-sected into additive single-gene pieces, and she could subsequently be re-assembled from them, just like the card game war can be reduced to an understanding of the context-free value of single cards.

The abstraction and reification of single gene effects is the foundation of standard theory. Heritability, h^2, is the fraction of the total trait variation among individuals that is owing to the additive genetic variance. However, nonadditive or interaction effects of genes contribute to the additive genetic variance (Hill et al. 2008) and do so to such an extent that "the existence of additive variance is not an indication that any of the genes act additively (i.e., show neither dominance or epistasis)" (Falconer and Mackay 1996, p. 126). Differently put, a fundamental parameter of individual selection theory, h^2, is itself composed of and derived from interactions.

In his studies of gene interactions, my student Dr. Charles J. Goodnight (1987, 1995) discovered "conversion," a process wherein variation caused by gene interactions is "converted" into additive variation by random genetic drift or by natural selection (chapter 5). In standard theory, drift and selec-tion together diminish additive genetic variation (h^2) and thus limit the adaptive process. Charles discovered that h^2 could increase even as overall genetic variation is decreasing, *but only when there are interactions between genes*. When one of two interacting genes becomes "fixed" by drift or selec-tion, the interaction effect between them is added to the main or "additive" effect of the remaining segregating gene. This conversion of interaction variation into additive variation can extend, rather than limit, the adaptive process, especially for fitness traits where interaction variation is predomi-nant. Conversion is a process that supplies h^2 to local populations, allowing a highly complex genetic system to be acted upon by individual selection *as though* its genetic pieces acted additively.

When different genes become fixed in different localities (demes) within a metapopulation, the effect of a single gene can change from place to place in both sign and magnitude as it interacts with different genetic back-grounds. Such genetic differences between populations contribute to group heritability, g^2. Thus, *both h^2 and g^2* have their origins in complex, interact-ing genetic systems. Whatever its role in Nature, studying group selection has enriched my understanding of adaptation through the discovery of new phenomena, such as conversion.

Lastly, interaction variance is more fundamental than additive genetic variance, in the same sense that particle physics is more fundamental than mechanics: drift and selection convert interaction variance into additive variance but additive variance cannot be converted into interaction variance. Interaction variance permits adaptive differentiation among populations, whenever a gene good for fitness on one genetic background becomes a gene bad for fitness in another background. In standard theory, a gene is assigned a constant, context-free effect when it arises by mutation — just like in war, where a card dealt maintains its value independent of other cards throughout the game. When gene interactions are studied, it is generally done by adding smaller, "second order" terms to an existing additive model. When a gene is defined as either good or bad when it arises by mutation, then its evolutionary trajectory is assured at birth. In contrast, if a gene can be good in one context but bad in another, it becomes much harder to justify leaving the genetics out of one's theory altogether as is done in some game theory models of evolution.

Interactions among the Levels of Selection

In most of the debate over group versus individual selection, discussion is focused on which force is stronger. The idea that the levels of selection might interact with one another has not really been acknowledged in this debate. One might expect that, if an adaptive change could be produced by either of two selection levels, then both levels acting together would be more effective at causing adaptation than either level acting alone. Experimental evidence shows that this is not always the case (Craig 1982; Goodnight 1985): selection at one level may interfere with a response to selection at another. When such interactions are incorporated into theoretical models, interference between the levels of selection is evident (Agrawal et al. 2001) and the unexpected experimental outcomes, like those of Craig or Goodnight, become easier to explain. One would also expect that, if an adaptive change could be produced by group selection, then group selection every generation should produce a larger response than group selection only once every two or three generations. Experimental evidence shows that this too is not the case (Wade and Goodnight 1991; chapter 8); group selection once in a while can produce effects as large as or larger than group selection all the time. Adding interactions into standard genetic models of evolution in metapopulations explains why this might be so.

Because different kinds of genetic variation affect the different levels of selection differently, selection at more than one level simultaneously can produce surprising and counter-intuitive results. For example, individual selection acting strongly on one gene, limits its own effectiveness when acting on other simultaneously selected genes in finite populations (called the *Hill-Robertson effect* or *selective interference*). As a result, strong individual selection is self-limiting. In contrast, group selection can be self-facilitating, in that selection on one trait can increase the genetic variation among groups for other traits and thereby enhance the opportunity for group selection to act later upon some other trait (Wade 2013; chapter 8).

Selection below the Level of the Individual

Like our studies of selection at levels above the individual, studies of selection levels *below* the individual have revealed aspects of variation, replication, and heredity that are different from, but enriching to, our understanding of individual selection. Contrary to the reductionist view that holds that higher level phenomena are merely the average effects of lower level processes, the study of genic selection has revealed *more complexity* at the lower level rather than less.

Gametic selection, also known as meiotic drive or segregation distortion, violates Mendel's Law of Segregation and is one of the strongest evolutionary forces known. Weak but consistent biases at meiosis or at fertilization, which favor one allele over another, can rapidly transform an entire population. This is why gametic selection was deemed important enough to put on our doctoral candidacy exam at Chicago.

On the surface, meiotic drive systems have all the hallmarks of the "selfish genes" hypothesized by reductionist extrapolation of Darwinian principles (Dawkins 1976): they appear to promote copies of themselves at the expense of everything else. However, all such systems have been found to be complex, multi-gene interactions and none have proven to be problems with Mendelian meiosis per se. Often they involve interactions between cells as well as between genes. The study of the heritability of individual traits, h^2, did not provide much insight into the nature of heredity at the lower level. If anything, it lent credence to the mistaken view that the hereditary objects favored by meiotic drive were single selfish genes. Studying individual heritability does not provide much insight into the features of group heritability, g^2, either.

Referring to pollen tube growth rates, Haldane (1932) stated that a "plant is at the mercy of its pollen." By this, he meant that a gene for rapid pollen tube growth with superior rates of fertilization would spread through a population even if it harmed other aspects of plant fitness. Haldane makes it clear that, in this conflict between levels of selection, he expects selection at the "lower level" in the hierarchy to win. In general, the efficacy of selection is believed to be inversely related to the level of selection, with lower levels uniformly stronger than higher ones. However, species-level selection has been discussed as a means for restricting meiotic drive of the sex chromosomes, a selection process three levels lower (Jaenike 2001).

A process similar to Haldane's biased fertilization occurs in animals, where it is called "cryptic female choice." Here, selection occurs among sperm of different males after mating, but prior to fertilization, within the sperm storage organs of multiply inseminated females (Eberhard l996; Shuster and Wade 2003). It is "cryptic" in the sense that it is difficult to correlate fertilization success with observations of courtship or male mating behavior. Pollen competition is similarly cryptic in that it is not a simple function of pollinator visitation rates or pollen load. In many of these instances, careful study has shown that fitness is not a property of the male gamete per se, but a result instead of complex interactions between male gametes or between them and female genotypes.

In short, interactions appear to be ubiquitous even for phenomena associated with selection below the level of the individual.

Conflict between Genomes

Conflict between genomes within individuals, such as that between genes in organelles and genes in the cell nucleus, has many of the same evolutionary genetic properties as meiotic drive or gametic selection. For example, the mitochondrial defect of cytoplasmic male sterility (CMS) in some plant species (McCauley and Taylor 1997) results in the overproduction of seeds and an absence of pollen in plants with the CMS mitochondrial type. Since mitochondria tend to be inherited maternally, passing only from mother to offspring, producing an excess of seeds favors the CMS mitochondrial type. CMS plants have higher seed fitness but produce no pollen. Thus, the advantage to mitochondria via the seed imposes a severe cost to total plant fitness through the loss of male reproductive function. Nuclear genes that restore pollen function and curtail excess seed production are overrepresented in the pollen and consequently are favored at fertilization, which requires both ovule and pollen. Thus, two types of gamete selection are acting in this system, one favoring the mitochondrial CMS gene and one favoring nuclear restorers of pollen function (Jacobs and Wade 2003). Each is the result of a complex process of interaction between genes in the nuclear and mitochondrial genomes. Like the case of meiotic drive above, some have proposed that a third level of selection, group selection, plays a role in maintaining the simultaneous polymorphism of mitochondrial and nuclear genes (McCauley and Taylor 1997).

Selection among Groups

Kin selection or selection among groups of relatives (kin) is the basis for the evolution of sociality. Kin groups can be single families or groups of more distant relatives. It is a combination of two levels of selection: selection between individuals within kin groups and selection among kin groups (Wade 1980 a and d). Early mathematical modelers (Williams and Williams 1957, p. 32) recognized kin selection as a multilevel selection problem: "we use favorable between-group selection to balance unfavorable within-group selection but our groups are sibships." However, in later work, particularly G. C. Williams' book (1966) *Adaptation and Natural Selection*, and W. D. Hamilton's (1964 a and b) seminal papers, "The genetical evolution of social behavior" I and II, evolutionary biologists found it preferable to understand the evolution of social behavior as a by-product of selection among self-interested individuals. Today, kin selection is described as selection among individuals to help genetic relatives (e.g., Griffin and West 2002), instead of as selection among kin-groups.

Classically, group selection has meant selection among local populations (demes; Wright 1931). It depends upon variation among populations in fitness, so that the fitness of some populations is greater or lesser than that of others. Population fitness manifests itself as differential extinction, colonization, or dispersion, or as the differential fusion and fission of groups. Despite Levins' conceptual work on metapopulations, it can be difficult to establish that groups in Nature have the properties of replication, variation in fitness, and heredity necessary for a selection process. Groups can be formed in many different ways: (1) by the fission or fragmentation of larger groups into smaller ones; (2) by colonization, with or without the mixing of individuals from different source populations; and (3) by the fusion of smaller groups into larger ones. Hence, at the group level, replication can involve one or a multiplicity of "parent" populations instead of the one- or two-parent systems on which individual selection is based. The process of group replication determines, in large part, whether biologically significant among-group genetic variation can exist (Wade 1978a; Wade 1996). Furthermore, once formed, a group may persist for a long time and exchange members with other groups through migration. This mixing between groups reduces g^2 (group heritability) because groups exchanging migrants necessarily become more genetically similar to one another.

If groups do have the requisite properties for selection, it is still not clear what, if any, category of adaptations or patterns in Nature can be better explained by group selection than by individual selection. Williams (1966) argued that adaptation is an onerous concept to demonstrate and proposed a "principle of parsimony," wherein, if an adaptation can be explained by individual selection, then there is no need to invoke group selection or any other level of selection. For this reason, much of the group selection controversy has focused on adaptations that are good for the group but harmful for the individual. That is, group selection has been reserved by many as an explanation for only those cases wherein Darwinian natural selection fails as an explanation.

Darwin (1859, p. 106) himself used a similar approach when he considered the evolution of the sterile castes of the bees, ants, and wasps. He saw their existence as "potentially annihilating" for his theory of individual selection, because sterile individuals do not leave offspring to future generations. He proposed selection among colonies (a type of kin selection) as the mechanism for the evolution of sterile castes:

> This difficulty, though appearing insuperable, is lessened, or, as I believe, disappears, when it is remembered that selection may be applied to the family, as well as to the individual, and thus may gain the desired end. Thus, a well-flavored vegetable is cooked, and the individual is destroyed; but the horticulturalist sows seeds of the same stock, and confidently expects to get nearly the same variety; breeders of cattle wish the flesh and fat to be well marbled together; the animal has been slaughtered, but the breeder goes with confidence to the same family. . . . Thus I believe it has been with social insects; a slight modification of structure, or instinct, correlated with the sterile condition of certain members of the community, has been advantageous to the community: consequently the fertile males and females of the community flourished, and transmitted to their fertile offspring a tendency to produce sterile members having the same modification.

The group-selection argument among evolutionary biologists is analogous to those arguments in human affairs, where the "good of the individual" is pitted against the "good of the group." For example, when a communicable disease is common and a large fraction of the population is susceptible, vaccination is good for the individual. After a period of successful vaccinations, however, an unvaccinated individual is protected from disease by "herd immunity," because most others have been vaccinated and are not disease transmitters. Indeed, the risk to an individual of disease or injury from the vaccine itself may be perceived to be greater than his/her risk of contracting the disease, as has happened with whooping cough or diphtheria in some communities in the US and the UK. Once the fraction of a population vaccinated is high enough (a fraction that varies considerably among diseases on the basis of their ease of transmission), an individual can benefit from the public good of herd immunity without contributing to it. Anti-smoking laws are another example where individual freedom to smoke may be curtailed in light of the health hazard to others of smoking in confined places. Likewise, freedom of speech and civil disobedience both can put an individual in conflict with society at large. Just as the study of group selection has been used to improve egg laying and animal welfare in hens (Muir 1996; Wade et al. 2010), maybe understanding group selection can enlighten us in one or more of these aspects of human affairs.

It is important to recognize too that concepts from human affairs have been imported into the scientific discussion of group selection. Most notably, a tendency to "cheat the system" or to be a "free rider" on a public good or group benefit has been described as an inherent property of "selfish" genes (Dawkins 1976). Gene self-interest frequently arises in the discussion of formal evolutionary models (e.g., Wade and Breden 1980). Most often, the mutational origin of selfish individuals or cheaters within groups is used as an argument *against* the efficacy of group selection to produce social adaptations in the first place. In fact, it is a widely held but mistaken belief that social groups are vulnerable to selfish cheaters (Van Dyken and Wade 2010).

The case of an adaptation that is good for the individual but harmful for its group has received much less theoretical or empirical attention, even though it is the most common situation in the domestication of plants and animals. Some selection experiments studying this kind of group-harmful trait have been carried out in both animals and plants. These include our laboratory research on the evolution of cannibalism in flour beetles (Wade 1980a); our field work (Breden and Wade 1989) on cannibalism in the willow leaf beetle, *Plagiodera versicolora*; Goodnight's (1985) experiments on leaf area in plants; Muir's (1996) experiments on egg laying in hens; and those of Ellen and collaborators (2008) on social mortality in hens. The results of these studies leave little doubt that artificial group selection can curb the evolution of traits good for the individual but harmful to the group. The question is: can such traits be curbed by group selection in Nature?

Wright's Shifting Balance Theory

The foundation for a formal theory of multilevel selection can be found in Wright's Shifting Balance Theory (Wright 1931, 1978), one of the most comprehensive evolutionary theories of the past century. Wright's theory emphasized gene interactions good for the individual that spread by group selection via differential dispersion (which Wright called interdemic selection). In Wright's theory, interdemic selection favored traits good for the individual but difficult to achieve by natural selection alone owing to the ubiquity of gene interactions and the context-dependence of gene effects. Wright's theory has been criticized (e.g., Coyne et al. 1997, 2000) for a lack of empirical evidence from natural populations. It has been labeled superfluous and unnecessary by the parsimonious claim that the standard one-gene-at-a-time model is sufficient to explain most adaptations. Nevertheless, experimental evidence from our laboratory (chapter 9) has illustrated the efficiency of Wright's group selection. And, ancillary experiments into the genetic basis of the response to interdemic selection indicate that gene-gene and genotype-genotype interactions are involved, just as Wright proposed. Conversely, our experimental studies estimating both within and among-population genetic variation for fitness have shown that the genetic variation among our populations is too large to be accounted for by standard additive theory. Ecological studies of natural metapopulations have shown that the processes of local extinction, colonization, and differential dispersion, the processes on which Wright's theory was founded, are common in Nature.

In the chapters that follow, I will discuss the development of our experimental, theoretical, and field studies and their linkages to one another. In chapter 2, for clarity, I provide a technical definition of *group selection*, in relation to other selection processes. There, I discuss group selection in the context of simple population genetic models and illustrate its relationship to individual selection, the Darwinian process of standard evolutionary theory. The conceptual background behind my dissertation research, "An experimental study of group selection" (Wade 1977), and the surprising findings from it are the subject of chapter 3. Those findings raised more questions than they answered and stimulated a cascade of both theoretical and empirical research (chapters 4 and 5) designed to place the early results on a more rigorous empirical and theoretical footing.

In chapter 6, I will discuss the conceptual relationships between kin and group selection, emphasizing the advantages of the multilevel selection approach to the evolution of social behavior. I will also offer a critique of the inclusive fitness approach more commonly used to study behavior evolution. In chapter 7, I discuss laboratory experiments testing the predictions of the kin selection models, and in chapter 8, the field experiments on kin selection that we conducted with willow leaf beetles, *Plagiodera versicolora*. These field studies allowed me to calibrate results from the flour beetle laboratory model against Nature. Our experimental studies of Sewall Wright's Shifting Balance Theory are discussed in chapter 9. Finally, in chapter 10, I discuss the unique signatures of polymorphism and divergence that group selection leaves on gene sequences and the open questions that I hope will stimulate future research on adaptation in metapopulations.

2   What Is Group Selection?

Introduction

The existence and efficacy of group selection have been debated in two different contexts. One debate centers on the existence of group adaptations. Individuals are deconstructed into their component adaptations and, for each such trait, one asks, "Who benefits?" If a trait benefits individuals, then it evolved as an adaptation for individuals by individual selection. If it is a trait that benefits groups, then it evolved as an adaptation for groups by group selection (see also chapter 3). The genetic basis of a response to individual or group selection is not an issue, because the focus is on existing adaptations evolved in the past and shared by all extant individuals or groups. Genetic variation and distinctions between h^2 and g^2 are not relevant.

The second debate is about the nature of the evolutionary process rather than the adaptations produced by it. There have been several theoretical approaches to this debate about the effectiveness of group selection relative to individual selection and how often, if ever, Nature resides in that portion of parameter space where group selection operates. One mathematical approach used in this debate is called *multilevel selection*—where two kinds of selection operate simultaneously: selection between individuals within groups and selection between groups. Each type of selection is modeled explicitly. In another approach, called *inclusive-*

fitness theory, the levels of selection are not explicitly modeled. For simple genetic models, there is an acknowledged equivalence between multilevel selection and inclusive fitness theory (e.g., Wade 1980d; Keller 1999). Nevertheless, some inclusive fitness adherents consider the multilevel selection approach "irrelevant" or "useless" for understanding Nature (e.g., Wild et al. 2009), despite its utility in agricultural practice (Wade et al. 2010) and its effectiveness for developing experimental protocols. My approach to evolutionary genetics is based on multilevel selection with its roots in quantitative genetics and animal breeding. We have used that approach to design experiments that have revealed some of the unique features of group selection.

I will discuss briefly both debates before defining what I mean by *group selection*. The two debates are isolated in the literature because one tends not to reference the other (Winther, Dimond, and Wade 2013).

The Group Selection Debates

Darwin's process of evolution by natural selection accounts for the "fit" of organisms to their environment. It explains trait function(s) in relation to increasing individual survival and reproduction. An adaptation is any trait that enhances, or at one time enhanced, the net survival and reproduction of the individuals possessing it. It is natural to ask whether or not there are traits that cannot be explained by Darwin's paradigm for this would be a weakness of his theory. It is also interesting to ask whether other levels of selection in the biological hierarchy might lead to the same fit of organisms to their environments. Let's consider the evolution of two traits that Darwin considered a weakness for his theory, the sterile castes of the bees, ants, and wasps, and the sex ratio of a population.

Sterile Castes

Darwin thought that the sterile reproductive castes in the social insects could be potentially "annihilating" for his theory. Although caste in the bees, ants, and wasps can be a highly specialized adaptation for defense, food storage, or brood care, sterile individuals do not leave descendants. Darwin argued that caste was a trait that benefits the colony, instead of the individual. He proposed selection among colonies or families—a type of group selection—as the process causing caste evolution. Darwin used the analogy of a breeder selecting fruit or meat for its flavor to illustrate his point. Although a particular animal or fruit is destroyed in the tasting, the breeder returns to its family for breeding stock. As a result, the family prospers despite the sacrifice of the individual. Similarly, modern breeders of dairy cattle and laying hens choose sires on the basis of the performance of female relatives, thereby combining individual selection of females with family selection of males.

In the "who benefits" debate, Darwin's remarks have been reinterpreted as a type of individual selection (Ruse 1980), wherein caste is interpreted as an *adaptation for* the queen, bestowing upon her a reproductive or survival benefit relative to other queens. Here, caste is not the result of group selection because it is not an *adaptation for* the group. Instead, it is seen as an *adaptation for* the queen and therefore evolved by individual (i.e., queen) selection. The unsatisfying equivalence of queen and colony is not addressed in this argument.

Since the 1960s, the "who benefits" question has become a *gene benefits* issue, what is known as the *gene's eye view* of evolution (Williams 1966; Dawkins 1976). An individual made sterile by a *gene for* sterility may have no fitness at all by the typical route of producing offspring. However, if copies of the same gene exist in its reproducing relatives, then it can spread *indirectly* by helping kin reproduce. (It requires a gene that does not completely determine whether its carrier is sterile or fertile and the same gene must enhance the tendency of its bearer to assist relatives.) In this scenario, a gene has two ways to achieve genetic fitness: (1) directly, by the reproduction of its bearers; or (2) indirectly, by enhancing the reproduction of relatives bearing copies of the gene. In the calculus of inclusive fitness theory, a *gene for caste* loses direct genetic fitness owing to sterility, but more than compensates for this loss by promoting copies of itself that reside in reproducing genetic relatives. In this way, a *gene for sterility* spreads through a population via individuals helping their relatives. *Is such selection individual selection as asserted by Ruse, gene selection as claimed by Williams and Dawkins, or group selection (i.e., among-family selection) as proposed by Darwin?*

Sex Ratio

Sex ratio was a trait that so puzzled Darwin (1871) that he abandoned its solution: "I formerly thought that when a tendency to produce the two sexes in equal numbers was advantageous to the species, it would follow from natural selection, but I now see that the whole problem is so intricate that it is safer to leave its solution to the future." Frequency-dependent individual selection favors a balanced sex ratio (Fisher 1930). The evolutionary logic is based on the fact that, because each offspring has one mother and one father, the majority sex must have a lower average fitness than the minority sex. Although many taxa have equal or balanced sex ratios, some have sex ratios that are female-biased, especially the bees, ants, and wasps.

The "who benefits" explanation for female-biased sex ratios is based on the unusual chromosomal arrangement shared by bees, ants, and wasps (and a handful of other organisms): females have two copies of every gene, but males have only one, inherited from their mother. As a result, the average female worker is genetically related to her queen mother by ½, to a brother by ¼, but to another of her sisters by ¾. For the queen's fitness, producing a daughter is genetically equivalent to producing a son, since each has half of her genes. As a result, the queen benefits by the standard 1:1 sex ratio. A sterile female worker, however, should favor sisters over brothers, since producing a sister is three-fold better than a brother. Here, colony sex ratio could be an *adaptation for* both the queen and the sterile female workers. A female-biased sex ratio represents an evolutionary compromise between the genetic interests of the two parties, queen and worker, rearing the brood. This theory can explain any specific degree of female-bias depending upon the degree of control that the queen or her workers, respectively, have over the brood sex ratio. Queens "control" brood sex ratio by differentially fertilizing eggs—with fertilized eggs developing into daughters, while unfertilized eggs become sons. The sterile workers "control" sex ratio by differentially provisioning males and females during development. It is admittedly hard to falsify a theory that can explain *any* sex ratio observation.

In multilevel selection theory, female-biased sex ratios (Colwell 1981) evolve because groups with more females grow faster than groups with equal sex ratios. These larger groups disperse more colonists—that is, fertilized females, to establish new populations than smaller groups with fewer colonizing females. Within groups, however, females with equal brood sex ratio have higher fitness than females with daughter-biased broods. Whether or not a female-biased sex ratio evolves thus depends upon the strength of individual selection within groups relative to the strength of among-group selection. (There are more complicated theoretical scenarios involving three different kinds of selection affecting the population sex ratio [cf. Jacobs and Wade 2003; Wade, Shuster, and Demuth 2003]).

The modern debate over *group adaptations* began with Wynne-Edwards' (1962, 1965) proposal that communication between individuals by display behavior was a group adaptation evolved by group selection. He argued that the adaptive function of communication was to allow group members to assess local density. In response to high density, members could coordinately limit their reproduction thereby staving off resource depletion and maximizing group persistence. His views are often referred to as "naïve group selection" (Wilson and Sober 1994; Pianka 2011, p. 121). An alternative explanation (Wiens 1966) was that natural selection optimizes the difference between reproduction and viability; it does not maximize reproduction. Lack (1947, 1948) had shown that birds that lay fewer eggs when resources are scarce leave more young than those that lay the maximum clutch of eggs every year, because the latter lose entire clutches to starvation in years of low resources. The optimal trade-off between clutch size and fledged young is an intermediate number of eggs, not the maximal number. In the field of life-history evolution, it is widely recognized that such trade-offs between the viability and reproductive components of fitness are common in Nature.

While few of Wynne-Edwards' claims have been definitively refuted or supported by data, recent multilevel selection models have shown that communication and reproductive limitation can evolve by group selection (Werfel and Bar-Yam 2004). It requires two kinds of genetic variation: (1) variation among consumers in the tendency to produce a signal when they discover a patch of resource; and (2) variation among consumers in the tendency to reduce reproduction when they receive a signal. Signaling, reproduction-limiting consumers invade nonsignaling, freely reproducing populations. Even though nonsignaling, freely reproducing individuals enjoy a local fitness advantage whenever they compete with the signalers, the former more frequently exhaust local resources and become locally extinct, as Wynne Edwards suggested.

Given the expected evolutionary vulnerability of cooperative social systems to "cheaters" (Keller 1999; Ghoul et al. 2014), it is surprising that this system cannot be invaded by either of two different kinds of "social cheaters." One kind of cheater is silent; it does not signal a resource discovery, but it does limit its own reproduction. The other kind signals, but does not limit its own reproduction. Neither selfish individual strategy spreads through the population (Werfel and Bar-Yam 2004)–each is eliminated by local resource depletion and extinction, the same processes that favored the evolution of communication and reproductive limitation in the first place. This model illustrates that, in any particular case, the "who benefits" question can be difficult to answer because benefits vary with time and have to be considered at both a local and global scale.

Practitioners of the "who benefits" or "adaptation for" perspective tend to privilege individual selection over group selection as an explanation for adaptations. That is, explanations rooted in individual or gene selfishness are preferred over those based on cooperation. This view is supported by the mathematical models of inclusive fitness where group context is treated as though it were a component of individual fitness. In this view, the cooperatively integrated colony or society is an illusion. Colony integration is an evolutionary equilibrium that arises as its inherently selfish members evolve to a standstill while pursuing competing reproductive interests (Dawkins and Krebs 1978; Dawkins 1982; Williams 1988; Queller et al. 1993). Williams (1988, p. 385) put it succinctly: "our modern concept of natural selection . . . can honestly be described as a process for maximizing short-sighted selfishness." In contrast, the multilevel selection perspective examines both selection processes simultaneously and tries to quantify the relative role of each process on adaptation.

Clearly, separating an integrated biological group into its parts is at least as difficult as deconstructing a complex machine into its parts. But, while no one would take a car to a mechanic to punish the carburetor for cheating on the distributor, this is the preferred evolutionary interpretation of groups in Nature. Indeed, Rice (2013, p. 217) claims that, "Because there are so many ways in which some parts of all. . . . genomes can evolve to gain a reproductive advantage at the expense of other parts, the prevalence of genomic conflict is universal, and it influences all aspects of genetic form and function." Although auto parts are subject to engineering trade-offs, they are in no sense in competition with one another; the carburetor is not in an evolutionary game where it can win at the expense of the distributor. We know that the two parts have been deliberately engineered to work together to produce a functioning vehicle. The biological problem is more difficult because, for any complex group we observe in Nature, we can tell two plausible stories about the origin of its cohesion and the interactions of its members. Under one hypothesis, the parts were deliberated engineered to work together by selection acting among groups—that is, multilevel selection. Under the other hypothesis, each part was engineered separately to seek advantage over the other parts by individual selection. In this latter view, a "cooperative group" exists only in appearance. It actually represents a competitive stalemate, better explained by nepotism, manipulation, or reciprocity. The group is a *selfish herd* (Hamilton 1971), not a cooperative herd. Should we accept one hypothesis over the other?

The Debate over Group Selection for Individual Traits

There is also some debate over the role of multilevel selection in the evolution of individual adaptations. This is surprising because the premise of this debate would appear to be incompatible with that of the debate discussed above. The important feature that distinguishes this debate from the previous one is that it is a process-centered rather than an outcome-centered debate. Lastly, this debate tends to be focused on the genetics underlying traits, unlike the adaptation-centered debate above.

This debate began with R. A. Fisher and S. Wright, two of the founders of the field of theoretical evolutionary genetics. Wright's analysis of domestication of shorthorn cattle for the US Department of Agriculture (USDA) showed that, although each farmer selected the best of his local stock to breed, the transformation of the breed occurred by a different process. Large changes in the breed occurred when a great bull appeared by chance on one farm and its semen was exported to the farms of other breeders desiring the traits of that bull. Just as Darwin used the artificial selection practiced by breeders as the basis for his theory of natural selection, Wright used this analysis of change in the characteristics of short-horn cattle as the basis for his own theory of evolution, the Shifting Balance Theory.

Wright proposed that favorable gene combinations arose in small, local populations (*demes*) by a combination of natural selection within each deme (*mass selection*) and random genetic drift (sampling variation in gene frequency in small populations). Whenever a local population, by a combination of mass selection and chance, arrived at a favorable gene combination conferring higher average fitness, that population would grow faster and to a larger size than other demes. The high local density of individuals would lead to emigrants from that population into other local populations with less favorable gene combinations. In this way, a favorable gene combination arising locally would disperse outward from its place of local origin throughout the species. That is, group or interdemic selection occurred as some groups dispersed more migrants than others. The parallels between Wright's Shifting Balance Theory and the processes transforming short-horn cattle are unmistakable.

Fisher, on the other hand, believed that natural selection was sufficient for adaptive evolution without recourse to random genetic drift. He pointed to evidence that selection, not chance, caused fluctuations in gene frequency even in large populations — evidence that he considered "fatal" to the "Sewall Wright Effect" (Fisher and Ford 1950, p. 117).

These two different viewpoints emerged from the different ways that Fisher and Wright treated complex gene interactions in their respective theories. Wright believed that gene interactions were ubiquitous and that the sheer multiplicity of possible genotypes guaranteed that gene combinations would vary by sampling among the many, small, local demes of a metapopulation. A gene that improved fitness in one deme might well lower it in another deme with a different genetic background. Each deme, like each farm, was a small experiment in adaptive evolution, more or less independent of other experiments. Group selection was necessary to spread an adaptive discovery arising by chance and selection in one deme to other demes across a species. Because organisms were more than the sum of their genetic parts, Wright was convinced that Nature required a mechanism for selecting directly among gene combinations.

In contrast, Fisher viewed natural populations as very large and so interconnected by migration that a species was a single evolving entity. Although gene interactions were common, they could be averaged to determine the effect of a single gene on fitness. Fisher was convinced that a gene's "main effect" on fitness (i.e., its grand average in all possible combinations) determined its evolutionary fate; no "extra" interdemic selection was necessary and random drift across the entire population of a species was negligible. Organisms were more than the sum of their genetic parts, but Nature built up genetically complex individuals one gene at a time. (The arguments and evidence for each side in this debate are discussed at length in Coyne et al. [1997], Wade and Goodnight [1998], Goodnight and Wade [2000], and Coyne et al. [2000].) Unlike the "who benefits" debate, where the genetic basis of traits is incidental or ignored, in this process debate, genetics is of the essence.

To better understand how these controversies influenced decades of research activity in the Wade lab, I next discuss the different levels of selection. Following that, I turn to the issue of how different genetic models affect each level of selection. That is, I will formally separate multilevel selection from the sources of genetic variation that underlie the traits that are the targets of selection. This method of separating selection from the genetic response to selection has long been important in the formal analysis of phenotypic evolution (Fisher 1930; Lande and Arnold 1983; Arnold and Wade 1984; Arnold and Halliday 1988). This approach allows me to distinguish the process of group selection from individual selection as well as to illustrate the role that genetics plays (complex versus simple) in the adaptive response to each level of selection. Some of our more complex experimental protocols were directed toward understanding multilevel selection, while others were focused on understanding the genetic response to that selection. In this chapter and others, I use simple schematic diagrams illustrate specific features of experimental design relevant to testing the efficacy of group selection and to dissecting the genetic response to it.

Different Kinds of Selection

I will first discuss selection among individuals and contrast it with selection among families, before turning to selection among larger groups, such as demes. Family selection is a type of group selection that is common in animal breeding and may well be common in Nature. In many organisms, the family is the most important group context affecting an individual's fitness. In our field work on willow leaf beetles, we measured the strength of selection within and among families in Nature and used these estimates to calibrate our laboratory experimental studies with flour beetles (chapter 8). The contrast between family and individual selection serves to illustrate the distinction between the properties of an individual and the properties of a family. I will show how maternal genetic effects that influence offspring fitness result in a very common type of family selection. Importantly, multilevel selection theory shows us that family selection leaves its own distinctive signature on the DNA of a population, making it possible to use molecular genomics to see the past action of family or group selection. In fact, we can see the theoretically predicted patterns left by family selection on the genomes of flies, mice, and humans (Demuth and Wade 2007; Cruickshank and Wade 2008; Wade et al. 2009). We have generalized multilevel theory beyond family selection, predicting the DNA sequence patterns of future sociogenomic studies of bees, ants, and wasps (Linksvayer and Wade 2009; Van Dyken and Wade 2012). When two alternative hypotheses can explain the same adaptation, it is important that they make other contrasting predictions so that we can gather evidence to distinguish between them. One of the most valuable attributes of multilevel selection theory lies in its predictions regarding patterns of genetic variation within populations and between populations or species. Importantly, these predictions allow group selection to be distinguished from individual selection with DNA data.

Variations in Survival and Reproduction

In Figure 2.1 (upper panel), I show a population of N-pairs of parents with their trait differences indicated by shading. The families are numbered 1 through N. One of these families, call it the i-th family, may differ from other families in the number of offspring the female parent produces (F_i) or in the average viability of those offspring (V_i). The *relative fecundity* fitness of this family, w_i, is (F_i/F.), where F. is the average fecundity across all females (i.e., $\Sigma_i F_i/N$). Similarly, the *relative viability* of offspring in the i-th family, v_i, is (V_i/V.), where V. is the average viability across all families (i.e., $\Sigma_i V_i/N$). Variations from one family to the next in w and v cause natural selection and the two values may be associated. Larger families might have healthier, more viable offspring (a positive association between F and V). Alternatively, larger families might have lower viability, owing to sibling competition for scarce parental resources (a negative association between F and V).

Experimental manipulations are necessary to dissect the causes of the variations in w and v among families. For example, consider the question, "Whose genes affect offspring viability, the offspring's own genes or mom's genes?" Experimental dissection is necessary to separate the *indirect genetic effect* of the maternal genotype on offspring viability from the *direct effect* of an offspring's own genotype on its viability. Without cross-fostering experiments (or their statistical equivalent [Wolf and Cheverud 2012]), the genetic influences of mother and offspring are confounded. One might also ask, "What difference does it make whose genes are responsible for an offspring's death? If we count offspring viability as part of maternal fitness, don't we get the correct evolutionary answer?" Unfortunately, we do not (Wolf and Wade 2001); in fact, when both direct and indirect genetic effects are present, we may even misidentify *the direction* of evolution. It is a mistake to count offspring fitness as though it were an aspect of parental fitness in all but the very simplest cases. For the same number of deaths, direct selection based on the offspring's own genotype is twice as strong as indirect selection based on the mother's genotype (Wade 1998). It is important to know *why* families vary in offspring numbers and *why* they differ in viability to understand adaptive evolution.

In the lower panel of Figure 2.1, I illustrate a metapopulation consisting of N demes, each deme consisting of K males and K females. Just like

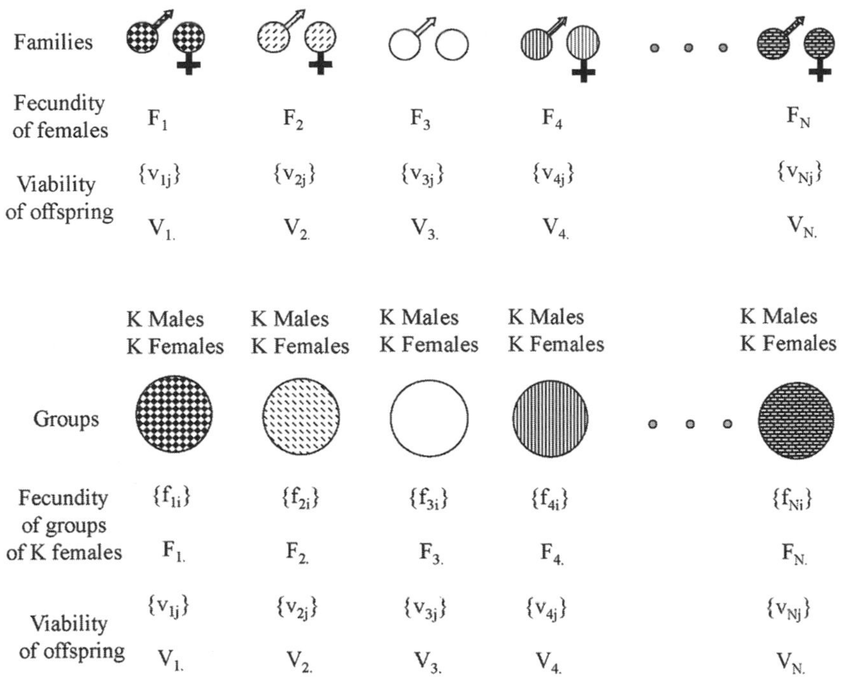

Families

Fecundity of females	F_1	F_2	F_3	F_4	F_N
Viability of offspring	$\{v_{1j}\}$ $V_{1.}$	$\{v_{2j}\}$ $V_{2.}$	$\{v_{3j}\}$ $V_{3.}$	$\{v_{4j}\}$ $V_{4.}$	$\{v_{Nj}\}$ $V_{N.}$

	K Males K Females	K Males K Females	K Males K Females	K Males K Females	K Males K Females
Groups					$\circ \;\circ\; \circ$
Fecundity of groups of K females	$\{f_{1i}\}$ $F_{1.}$	$\{f_{2i}\}$ $F_{2.}$	$\{f_{3i}\}$ $F_{3.}$	$\{f_{4i}\}$ $F_{4.}$	$\{f_{Ni}\}$ $F_{N.}$
Viability of offspring	$\{v_{1j}\}$ $V_{1.}$	$\{v_{2j}\}$ $V_{2.}$	$\{v_{3j}\}$ $V_{3.}$	$\{v_{4j}\}$ $V_{4.}$	$\{v_{Nj}\}$ $V_{N.}$

FIGURE 2.1 Upper panel: Pairs of individuals that differ from one another in fecundity of females (F_i) and the mean viability ($V_{i.}$) of the offspring of the i-th female (i.e., the mean of the set $\{v_{ij}\}$). Lower panel: Groups of K males and K females that differ from one another in the net fecundity of the females ($F_{i.}$) and the mean viability ($V_{i.}$) of the offspring of the i-th group of females.

the i-th family above, the i-th deme is different from the others as indicated by its shading. The K females in this group produce a number of offspring (F_i) whose average viability is V_i. The fecundity of group i, $w_{i.}$, *relative* to the metapopulation average, $F_{..}$, is ($F_i/F_{..}$). Similarly, the relative viability of the i-th group, $v_{i.}$, is ($V_i/V_{..}$) where $V_{..}$ is the average viability of all offspring across all groups. Variations from one group to the next in both $w_{.}$ and v can cause group selection and the association between $w_{.}$ and $v_{.}$ can be either positive, zero, or negative. Overall, individuals, families, and groups may differ in survival and reproduction and these differences represent an opportunity for selection. When demes have large differences in w and v, group selection can be strong.

Selection within and among Families

In Figure 2.2, I contrast individual and family selection for larger body size. In both schemes, larger individuals are selected for breeding while smaller individuals are discarded. Family selection (Figure 2.2 lower panel) is different from individual selection (Figure 2.2 upper panel) in that some small individuals are selected because their families have a high mean size and some large individuals are discarded because their families have a low mean size. That is, with family selection, an individual's fate, to be selected or not, depends not only on its own size but also on the average size of its siblings. Comparing individual and family selection, we can see that some of the same individuals are selected under either process. After selection, favored individuals are paired, randomly or otherwise, to mate and produce the next generation. The gain in size from selection can differ for the two selection methods depending upon the genetic causes of the individual variations in size.

Within-Family or Soft Selection

There is a third kind of selection, *within-family selection*, illustrated in Figure 2.3. Here, the largest male and the largest female from each family are selected for breeding. Note that one male and one female from the "white" family are selected for breeding under this type of selection (Figure 2.3) but none were selected from this family under *either* of the two previous types of selection. That is, this type of selection allows some of the smallest individuals to breed even though the goal of selection is to increase body size. Why would a breeder ever do this? Why would agricultural and dairy governing bodies *recommend* that breeders practice this kind of selection?

Individual selection and the pairing of selected individuals at random (Figure 2.2 upper panel) result in many pairings between members of the same family. This results in inbreeding, which can lead to a decline in growth rate, health, or fertility. Although strong individual selection leads to more rapid improvement than weaker selection, the inbreeding effect is greater with stronger selection because more individuals come from fewer families. Moreover, the problem of inbreeding is cumulative. With more generations of selection, the animals of today are descended from fewer and fewer of the original ancestors. Individuals in the selected population become more closely genetically related as selection is continued. Within-family selection is one of the most frequently recommended strategies for mitigating the

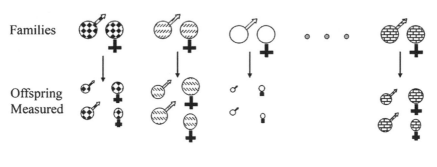

Individual Selection or **Hard Selection** for larger size:

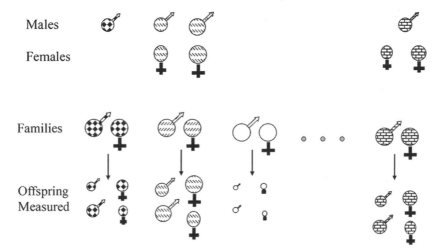

Among-Family Selection or **Group Selection** for larger size:

FIGURE 2.2 The contrast between individual (upper panel) and family selection (lower panel) for larger body size. Note that individuals and families differ from one another in body size. In this example, the experimenter causes individual and/or among family differences in viability (i.e., variations like those of v_i and V_j in Fig. 2.1) by the way in which he/she selects individuals to found the next generation. With individual selection (upper panel), the four largest males and the four largest females are selected, but with family selection (lower panel), two males and two females are selected from each of the two families with the largest *average* body size.

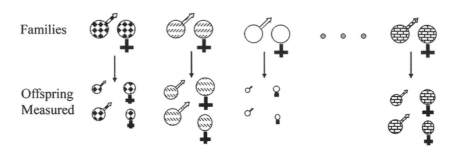

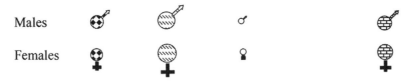

Within-Family Selection or **Soft Selection** for larger size:

FIGURE 2.3 Within-family or soft selection for larger body size, where the largest individuals in each family are selected for breeding. Note that the largest individuals from the families of small mean size are selected even though they are smaller than some of the individuals discarded from families of larger mean size. Breeders use this type of selection to mitigate inbreeding by loss of lineages owing to selection. Importantly, it reduces the variance in fitness among families toward zero. When Nature does this type of selection, it is called *soft selection* and the mechanism is the ecological regulation of density at the level of the family.

deleterious effects of inbreeding that attend individual selection. Within-family selection reduces the rate of inbreeding by one half. While it reduces the short-term selection response, it allows a greater long-term response. By reducing the among-family variation to zero, within-family selection maintains the largest number of maternal lineages.

Within-family or within-group selection is often called *soft selection* (Wade 1985b), because individuals are evaluated according to the local mean and not relative to the global mean. In contrast, selection where an individual is evaluated relative to all other individuals in the population is called *hard selection*. We might say that the white male here is a "big fish in a small-fish pond." He is selected for breeding when his *relative* size is measured by the local standard of his own family, but not when his *relative* size is measured by the global standard, which includes the members of other families. Note, with soft selection, more individuals are culled in large than in small families. A male and a female from a family of 50 constitute a much smaller fraction of that family than a male and a female from a family of 10.

We have three selection methods: individual, family, and within-family selection. How are they related to one another? How does each kind of selection interact with the underlying ecology and genetics of the trait targeted for selection (here body size) to produce an adaptive response to selection?

Simple Genetics and the Kinds of Selection

Consider a large randomly mating population and a simple, additive genetic model with only direct genetic effects. Let the viability of an individual increase by the amount s for each A allele in its genome. This is the simplest genetic model, where each individual has only a single trait, viability, which equals exactly the sum of its genetic parts. There is no genetic or ecological complexity.

Let the frequency of the A allele in the population be p, and the frequency of the alternative a allele be q. The genetic variation in the population is pq. Imagine that females mate many times, so that their eggs are fertilized by a representative frequency of A bearing sperm and a bearing sperm, resulting in the offspring frequencies within families given in Table 2.1. Averaged across all individuals (AA, Aa, and aa, in frequencies G_{AA}, G_{Aa}, and G_{aa}, respectively), mean individual fitness, W, is $(1 + 2sp)$. In this population, the total change of gene frequency caused by selection is

$$\Delta p = (pq)s/W. \tag{2.1}$$

How does the rate of evolution, measured by Δp, change if we consider *within-family* selection, evaluating individuals relative to the average via-

TABLE 2.1 Direct effect, s, on offspring viability with polyandry, i.e., many mates per female.

Mother's Genotype	Mother's Frequency	Offspring Genotypes			Mean Family Fitness
		AA	Aa	aa	
AA	G_{AA}	p 1 + 2s	q 1 + s	—	1 + (1 + p)s
Aa	G_{Aa}	p/2 1 + 2s	½ 1 + s	q/2 1	1 + (1/2 + p)s
Aa	G_{aa}	—	p 1 + s	q 1	1 + (0 + p)s
Mean Genotype Fitness		1 + 2s	1 + s	1	1 + 2ps

bility of their own family instead of relative to the global average viability? Here, the change in gene frequency is reduced to $\{(spq/2W) + (G_2s/8W)\}$. If we assume that selection is weak so that $G_2 \sim 2pq$ even after selection, then this expression simplifies

$$\Delta p = (3pq/4)s/W. \qquad (2.2)$$

The rate of evolution is reduced because we are now selecting on only the ¾ of the total genetic variation, pq, that is *within families*.

In this example, females mate many times. If each female mated only once, the rate of evolution would be further reduced because only half of the total genetic variance would be *within families*; moreover, only the *genotypic* portion of the genetic variance would be available for selection. (That is, there is genetic variation within aa × AA families, since all offspring are Aa heterozygotes. Yet, there is no selection at the genotypic level within these families since all offspring genotypes are identical.) Thus, the system of mating, polyandry or monogamy, influences the rate of evolution by within-family or soft selection because it affects the average amount of genetic variation within families. As shown in chapter 8, a genetic relationship between mating males and females — that is, inbreeding, decreases the amount of genetic variation within families and increases the amount between families, changing the balance between the two types of selection toward group selection.

If we restrict ourselves to *family selection* in Table 2.1, the rate of evolution changes again. Here, families of AA females have the highest average fitness, while families of aa females have the lowest average fitness. Like the small male of the "checkered" family in Figure 2.2, when we select whole families based on average family fitness, the *Aa* offspring of AA mothers enjoy a higher fitness than they would if evaluated solely on their own merit. Here, we find that the change in gene frequency equals $s([G_1/2] + [G_2/8] - (p^2/2])$. If we make the typical assumption that selection is weak, then $G_1 \sim p^2$ and $G_2 \sim 2pq$ after selection, and we find that, with only family selection,

$$\Delta p = (1pq/4)s/W. \qquad (2.3)$$

As a result of polyandry, only ¼ of the genetic variation, pq, is between families and available for family selection. The rest of the genetic variation $([3/4]pq)$ is within families, as we saw above when we calculated the rate of evolution by within-family selection. If females mated only once (monogamy), then family selection would become stronger since ½ of the genetic variation would be among families, so that $\Delta p \sim (1pq/2)s/W$. Clearly, the magnitudes of between-family selection and within-family selection are affected by the system of mating.

We can order our three types of selection with respect to the magnitude of evolutionary change:

Family Selection ≤ Within-Family or Soft Selection
≤ Individual or Hard Selection

Notice that, with either kind of random mating (monogamy or polyandry), what is called *individual selection* is the sum of among-family selection and within-family selection. When the genetics of viability is complex, this need not be the case (cf. Figure 2.3, for example). Furthermore, family selection with simple genetics is fairly weak because, just as the individual is nothing more than the sum of its genetic parts, the family is nothing more than the sum of its genetic individuals. The lack of genetic complexity allows us to establish a simple hierarchy of genetic variation: the variation among genes is twice as large as that among individuals (random samples of two genes), which in turn tends to be greater than that among families (a random sample of four or more genes, depending on the mating system).

Hard and Soft Selection Simultaneously

Instead of the alternatives of hard selection or soft selection, we can have any combination of them. This is often ecologically interpreted as density regulation occurring on a sliding scale between local (soft) and global (hard) owing to negative feedbacks on population growth rate. In Figure 2.4, I depict a metapopulation in which a fraction of the population, c_i, resides in the i-th deme and this deme has a frequency of the A allele equal to p_i. Individual selection *within groups* changes the allele frequency in the i-th deme from p_i to p_i': $\Delta p_i = sp_iq_i/W_i$. It is as though each deme were considered as a stand-alone population in the simpler case (see eq. [2.1]).

Does local selection change the relative population size of each deme? Does the i-th deme contain the same fraction, c_i, of the total population after selection as it did before selection? Or, does c_i change to some other value, c_i'? To answer these questions, we need to recognize that individuals are dying for two reasons. Some die for genetic reasons, while some die for

Selection in a Metapopulation

Before Selection:

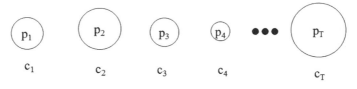

After Selection:

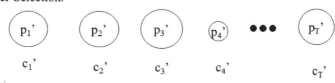

FIGURE 2.4 A genetically subdivided population or metapopulation consisting of T demes. The size of the i-th deme relative to all other demes before selection is c_i and the frequency of the A allele in this deme is p_i. After selection, the frequency in the i-th deme is p_i' and the size of the i-th deme relative to all other demes is c_i'. The value of c_i' relative to c_i depends upon whether density regulation is local (soft selection) or global (hard selection). (See text for discussion.)

ecological reasons. If there is strictly local density regulation and individuals die for local ecological reasons, then c_i does not change. Although some individuals die for genetic reasons, their deaths replace ecological deaths that would have happened anyway. If there is strictly global density regulation, then genetic deaths affect the size of local demes and c_i changes in proportion to the relative strength of selection in the i-th deme relative to all other demes.

Let the parameter b represent the fraction of hard selection and (1–b) the fraction of soft selection, so that

$$c_i' = (1-b)(c_i) + (b)(c_i)(W_i/W.). \tag{2.4}$$

Letting F_{ST} be Wright's fraction of the genetic variation among demes, the total change in gene frequency (after Wade 1985b) is

$$\Delta p = \Delta p_{within\ groups} + \Delta p_{between\ groups,} \tag{2.5a}$$
$$\Delta p = \{spq(1-F_{ST})/W\} + \{spq(2bF_{ST}])/W\}. \tag{2.5b}$$
$$\Delta p = spq(1-F_{ST}[1-2b])/W. \tag{2.5c}$$

When b is 1, there is only global density regulation—that is, strictly hard selection. When b is 0, there is only local density regulation—that is, strictly soft selection. In eq. (2.5b), the second term on the right-hand side is group or among-deme selection. It is easy to see that group selection is reduced by strong local density regulation (i.e., small b). As Wright argued, local selection must affect the mean fitness of local demes in order to create an opportunity for group selection.

Because an individual's contribution to the next generation is determined by both its genetic make-up and the *ecology* of density regulation, we can no longer equate individual selection with hard selection. In the model, we know that individuals live or die for independent genetic and ecological reasons. In Nature, we must do experiments to discover what fraction of the mortality is owing to genes and what fraction to ecology. Indeed, the ecology could also cause inbreeding ($f \neq 0$) if individuals mated locally before dispersing, further complicating the conception of a simple dichotomy between group and individual selection.

In summary, for this simplest of genetic models with ecology, there is no necessary order of the levels of selection with respect to the magnitude or strength of evolution. The ecology of density regulation can reverse the relative strengths of within-group individual selection and group selection. In the simple genetic models without ecology, we could argue that the distinctions among the different kinds of selection are semantic and hold little meaning. However, when even simple ecology is added to our genetic models, the distinctions become critical.

Not So Simple Genetics: Direct versus Indirect Genetic Effects

Now imagine a different kind of Mendelian gene, also with two alleles—one which has a *direct effect* on fitness, s, and an *indirect effect* on fitness, s_s. The indirect effect of this gene causes an individual to have an effect, s_s, on the fitness of other individuals in its family (or in its deme). Individuals with genes of this sort can be considered socially cooperative (when $s_s > 0$) or socially competitive (when $s_s < 0$). This is illustrated in Table 2.2. Notice that *all* individuals within a family have their fitness incremented by the same amount but that families differ from one another in average fitness because they have different numbers of social members.

For a gene with effects like this, the equations for within-family, among-family, and total selection are

Within-family selection: $\Delta p = (3pq/4)s/W$, $\qquad$ (2.6)

Among-family selection: $\Delta p = (1pq/4)(s_s + s)/W$, and $\qquad$ (2.7)

Total selection: $\Delta p = ([s_s/4] + s)pq/W$. $\qquad$ (2.8)

Notice that the indirect effect of this gene (s_s) does not influence within-family selection (eq.[2.6]). When $s_s < 0 < s$ and $(s_s + s) < 0$, this means that mean fitness, $W = 1 + 2p(s_s + s)$, declines as the A allele spreads to fixation. Mean fitness declines because the gene's positive direct effect, s, is offset by its negative indirect effect, s_s. This type of genetic scenario, where an individual benefits at the expense of its neighbors, may be the most common type of competitive interaction in plants and animals (Van Dyken and Wade 2012). When resources are scarce, the struggle for existence results in winner and losers. It is common for individuals of many plant species to achieve high yield by taking nutrients, like moisture and light, away from their neighbors. For plant breeders, ignoring these competitive interactions has consequences; individual selection for increased yield leads, instead, to a decrease in yield because high yielding individuals make nasty neighbors. Group selection is the only way to guarantee an increase in plant yield when such competitive interactions between individuals are common and have a genetic basis (Griffing 1967; Wade et al. 2010).

Such genetic interactions are ignored when within-family selection is recommended to animal breeders as a means to preserve genetic diversity and to avoid inbreeding depression. For conservation biologists, within-family selection is often recommended to minimize inbreeding depression in already genetically limited stock and to slow the rate of adaptation to cap-

TABLE 2.2 Direct effect, s, and indirect effect, s_s, on offspring viability with polyandry, i.e., many mates per female.

Mother's Genotype	Mother's Frequency	Offspring Genotypes			Mean Family Fitness
		AA	Aa	aa	
AA	G_{AA}	$\dfrac{p}{1+2s+(1+p)s_s}$	$\dfrac{q}{1+s+(1+p)s_s}$	—	$1+(1+p)(s+s_s)$
Aa	G_{Aa}	$\dfrac{p/2}{1+2s+(1/2+p)s_s}$	$\dfrac{1/2}{1+s+(1/2+p)s_s}$	$\dfrac{q/2}{1+(1/2+p)s_s}$	$1+(1/2+p)(s+s_s)$
aa	G_{aa}	—	$\dfrac{p}{1+s+ps_s}$	$\dfrac{q}{1+ps_s}$	$1+(0+p)(s+s_s)$
Mean Genotype Fitness		$1+2s+3p/2s_s+s_s/2$	$1+s+3p/2s_s+s_s/4$	$1+3p/2s_s$	$1+2p(s+s_s)$

tivity in re-introduction programs (e.g., Frankham 1995; Rosvall et al. 1998). As shown above, this recommended practice could well lead to an increase in competitive interactions (when $s_s < 0$), which will produce a decline in mean fitness, much like the inbreeding depression that within-family selection is designed to avoid.

When Breeders Turn to Group Selection

Breeders of domesticated plants and animals have identified three circumstances in which family selection results in a greater response than individual selection: (1) when the environmental component of phenotypic variation is greater than the genetic component, (2) when gene combinations are responsible for phenotypic variation among individuals, and (3) when social interactions or indirect genetic effects contribute to the phenotypic variation among individuals. All three cases refer to circumstances under which the heritability of a trait is low. As Crow and Aoki (1982, p. 2630) put it, "within-group selection depends mainly on the additive component of variance whereas between-group selection depends on the total genetic variance. Thus, we should expect that the lower the heritability of the trait the greater the relative effectiveness of group as opposed to individual selection."

Environmental Variation

When environmental variation is much greater than genetic variation, the family-mean is a better predictor of an individual's genotype than the individual's own phenotype, because the family-mean averages over the effects of environment on family members. Differently put, when h^2 is low, owing to the sensitivity of a trait to environmental variation, using phenotypic information from the individual and its family members permits the breeder to make a more reliable selection decision. A breeder has more confidence that families differ genetically from one another than he or she has that individuals differ genetically. I have illustrated this in Figure 2.5, where the black coloration depicts the genetic contribution to body size and the gray coloration, the environmental contribution, making the environmental contribution nine times greater than the genetic on average. If individuals were chosen on the sole basis of their own size, the middle member of each family would be chosen for breeding. As a result, the genetic composition

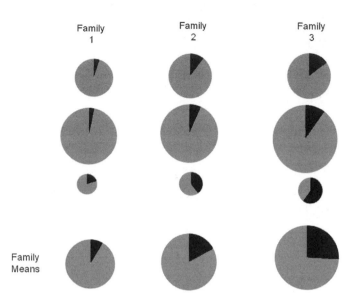

FIGURE 2.5 Genetic (black) and environmental (gray) effects on body size. See text for discussion.

of the population would be the same in the next generation as it was in the parents. However, if families were chosen on the basis of the family mean size, the large but random environmental variation would "average out" and the members of Family 3 would be chosen for breeding. In this illustration, family selection produces a greater increase in size in the next generation than individual selection.

Another, more common feature of environmental variation also makes individual selection a weaker evolutionary process than a combination of family selection and within-family selection. When there is Genotype × Environment Interaction (G × E), the effect of a gene on a trait may change in magnitude or in sign as the environment changes (Wade 2002; Wade 2014). My family sometimes introduced an analogous interaction into our card games when we played "upside-down war," where 2's were the high cards and aces were the low cards. This change in card value with game rules is analogous to change in a gene's effect with environment. With G × E, a gene that is good for fitness in one environment is a gene bad for fitness in another. Within a deme's local environment, the evolutionary game is still based on the simple, independent effects of genes. The problem is that the fundamental value of a gene (or a card) changes between environments (or between games). In this circumstance, a combination of family selection and within-family selection *always* produces a greater response than individual selection alone.

Gene Interactions

Gene interactions (G × G) become important when an individual's trait value depends on a particular gene combination, as opposed to being the simple sum of the effects of an individual's genes. Just like G × E, in this case, the effect of a gene on a trait changes in magnitude or in sign as the genetic background changes (Wade 2002; Wade 2014). The effect of G × G is to lower h^2 in a manner similar to either G × E or to large amounts of environmental variation. In fact, R. A. Fisher lumped the variation owing to G × G together with the environmental variation in order to have a single term representing the "non-heritable" sources of individual variation (Skipper 2002). But, G × G is non-heritable only from the viewpoint of strict individual selection. Genetic background itself is heritable at the level of the family or the group. Like G × E, when there is G × G a combination of family selection and within-family selection *always* produces a greater response than individual selection. When we discuss our experimental test of Wright's Shifting Balance Theory in chapter 9, we will see that his theory is essentially an assertion of this fact in regard to the evolution of fitness itself, a trait inordinately affected by gene interactions.

Social Interactions

We saw above that family selection is necessarily more effective than individual selection when there are genes with indirect or social effects (Drown and Wade 2014). When the social environment is caused by genes, it evolves in response to family selection. When the goal is to maximize yield (or fitness), family or group selection allows the breeder to select on the social context. For this reason, group selection for increased yield gives a positive response where individual selection gives a negative response.

In short, the levels of selection debate cannot be settled by asking "Who benefits?" (Lloyd 1988); the genetic and ecological basis of the response to selection must be considered. In the absence of interactions, individual selection is effective and sufficient. With interactions, either between genes, between genes and environments, or between individuals, it becomes much less so. Breeders have long recognized this and turned away from strict individual selection to combinations of family and group selection. Does Nature ignore interactions like much of classic theory suggests or does she behave more like animal and plant breeders?

In the next chapter (chapter 3), I discuss the intellectual climate of the group selection controversy at the time of my dissertation.

3   Group Selection in the 1970s

Introduction to Graduate Studies

A double major in mathematics and biology from Boston College secured my admission to the Department of Theoretical Biology at Chicago with four years of NIH training grant support. Like most graduate students, I struggled to learn sufficient background to devise a doctoral proposal from my readings, classes, and seminars. Retrospectively, my conducting a dissertation on the experimental study of group selection in the shadow of Williams' (1966) monograph may seem novel, bold, or contrarian. My dissertation proposal, however, was simply a research opportunity that emerged in the standard way from current conceptual controversy, field observations, and laboratory experiments.

Classes, seminars, and dissertations were focused on the tenets of the theory of island biogeography, which assumed that the membership of communities and species was distributed as a series of more or less isolated islands. An archipelago under this theory was a dynamic metapopulation, with frequent local extinction and colonization. Ecology and evolution at that time were not the conceptually distinct fields they are today. It was widely believed that the processes that maintained species diversity within ecological communities would bear a mathematical similarity to those that maintained genetic diversity within populations. So, ecologists were interested in new developments in evolutionary genetics and vice versa.

A key claim of island biogeographic theory was that community subdivision maintained biodiversity by permitting the coexistence of species that would otherwise go extinct. A single laboratory experiment supported this view. Using a metapopulation created from a grid of oranges, Huffaker (1958; Huffaker et al. 1963) had shown that a predatory and a herbivorous species of mite could stably coexist as long as the prey species dispersed to new, uncolonized oranges faster than the predator species. Faster colonization allowed the prey population to grow for a time unimpeded by predation. My experimental studies of group selection were a natural outgrowth of my exposure to island biogeography and to the research interests of my dissertation advisors. Dr. Thomas Park (1948, 1954) studied competition in experimental microcosms using flour beetles in the genus *Tribolium*, and Dr. Montgomery Slatkin (1974) was extending mathematical models of predator-prey interactions to discover the conditions that would permit regional coexistence between competing species. He (Maynard Smith and Slatkin 1973) had generalized Huffaker's experimental findings, extending existing predator-prey theory to subdivided metapopulations. Slatkin (1974) also showed in theory how two competing species — one a better competitor, the other a better disperser — could coexist in a metapopulation with a dynamic process much like that of Huffaker's predator-prey system. Understanding the coexistence of interacting species was viewed as key to understanding patterns of community diversity as well as genetic diversity.

In the summer of 1972, I used Slatkin's ideas in my first solo-teaching adventure. During college, I had been a summer counselor for two years with the Boston College Upward Bound Program for Boston public high school students. After my first year in graduate school, I became an instructor of biology. With the help of the students, I invented a predator-prey game based (loosely) on Slatkin's models. The class was divided into two teams, predator and prey, and each devised a strategy involving number of moves per turn and a reproductive rate at the end of a turn. Prey reproduced (i.e., brought team mates onto the board) if they survived; predators reproduced if they ate. The prey tended to be eliminated quickly, and the class goal after each game was to adjust the reproduction and dispersal parameters to allow coexistence for at least 10 turns. The moves took place on our "board," the desk-tops of the classroom. Whatever its educational merits, the predator-prey game was great fun, with students jumping from desk to desk at my commands, "Move!" or "Reproduce!" However, right in the middle of our exercise, the program director arrived with a site-visiting team from the Department of Education. Unfortunately for me, their evaluations placed an emphasis on "classroom decorum," and I was not invited back to teach the following summer, squashing my early experiment in what is now called *active learning*.

Searching for a Dissertation Topic

I had read Wynne-Edwards' (1964) article in *Scientific American,* "Population control in animals," as a college student. It stimulated my interest in social behavior to the point that I had quoted R. M. Yerkes, "One chimpanzee is no chimpanzee," in my graduate application essays. In graduate school, this same interest drew me to the controversy over group selection and its role in the evolution of social behavior.

The doctoral program at Chicago included courses on biological clocks from A. Winfree, developmental biology from S. Kauffman, mathematical modeling from M. Slatkin, circuit-theory models of the brain from J. Cowan, tropical ecology from D. Janzen and M. B. Lloyd, population ecology from T. Park, animal behavior from S. Altmann, population biology from R. Levins, and population genetics from R. Lewontin. In addition, we were required to take biochemistry, stochastic processes (Feller 1971, Volumes I and II), statistics, and prokaryotic and eukaryotic genetics (taught by Drs. M. and R. Esposito).

Weekly seminars were heavy with mathematical and conceptual content. Theoretical seminars typically ended with far-reaching speculative assertions, reigned in during discussion by counter-examples from Nature offered by the field biologists, Janzen, Lloyd, and Altmann, "the wet blankets of reality." The keen interest in theory was not unique to Chicago although few other places had such a concentration of theoreticians. The times were aptly characterized by Dr. David E. McCauley's phrase, "Every hallway conversation is another letter to the [American] Naturalist."

In class, the faculty tended to lecture about topics related to their current research with little or no introduction to the history of the problem under discussion. I found it difficult to understand whether one problem was more important than another and, if so, why. Park's Population Ecology course was markedly different; he laid out the historical origins of modern questions and the layers of empirical evidence from field and laboratory gathered by earlier researchers. He taught that the important questions were long-standing and that each generation of scientists tried to move them toward an answer. He emphasized that important problems sometimes "fell out of fashion" and could sit neglected and unresolved for decades.

In the graduate courses, you were expected to pick up the background necessary to understand the lectures on your own; some courses had syllabi, most did not. I developed background by reading the recent papers of my instructors. These included "Extinction" (Levins 1970), "Units of selection" (Lewontin 1970 b), "Is the gene the unit of selection?" (Franklin and Lewontin 1970; followed by Slatkin 1972), "Host plants as islands in evolutionary and contemporary time" (Janzen 1968, 1973), and "Selection in populations with overlapping generations I" (B. Charlesworth 1970). Prompted by Dr. D. B. Mertz, a former student of Park, I also read "Gambling for existence" (J. Reddingius 1971), which developed the view that a species could persist in the face of local extinctions if its members were dispersed across space. These papers, in combination with recent books, like *Adaptation and Natural Selection* (Williams 1966), *Evolution in Changing Environments* (Levins 1967), *The Theory of Island Biogeography* (MacArthur and E. O. Wilson 1967), *The Insect Societies* (E. O. Wilson 1971), and *An Introduction to Population Genetics Theory* (Crow and Kimura 1970), served as my introduction to population biology and evolutionary genetics. I learned the logic of Darwinian theory from Lewontin (1970) and read Darwin for the first time as a graduate teaching assistant in Slatkin's undergraduate course on evolution, where it was a required text.

Papers and seminars promulgated the view of ecosystems as islands, with the membership of a species distributed in small patches separated by territory inhospitable to migrants. Janzen was a particular advocate of this view—for example, "Host plants as islands in evolutionary and contemporary time" (Janzen 1968, 1973). The field studies by D. Simberloff and E. O. Wilson ("Experimental zoogeography of islands: a two-year record of colonization," 1970) tested predictions of island biogeographic theory, fumigating whole mangrove islands and observing subsequent patterns of recolonization (MacArthur and E. O. Wilson 1967). As a student, I interpreted this work as the ecological counterpart to Lewontin's evolutionary genetic analysis in "Selection for colonizing ability" (Lewontin 1965).

This view of species as metapopulations permeated all research and discussion. It appeared to offer very favorable conditions for group selection in Nature. However, many, most notably Williams (1966), argued on the grounds of parsimony against invoking group selection to explain adaptations, a position endorsed by Lewontin in his review of Williams' book. In contrast, Levins (1962, p. 361), was explicit about group selection as a process for maximizing population fitness: "The present study is an attempt to explore systematically the relationship between environmental heterogeneity and the fitness of populations. For each pattern of environment examined, we will determine which population characteristics would provide the maximum fitness, where fitness is defined in such a way that interpopulation selection [group selection] would be expected to change a species toward the optimum (maximum fitness) structure." When these theoretical models were synthesized in Levin's 1967 monograph, however, all such explicit references to "interpopulation selection" were omitted. Such was the shadow thrown by Williams (1966) over group selection.

Life-history theory was another area of active research that capitalized on group selection as a mathematical approach. Here, two or more populations were set up, each governed by a different birth and death rate. Next, the value of r, the intrinsic rate of population growth, was determined for each population on the basis of its demographic pattern. The population growing at the higher r would displace all other populations. From this, it was inferred that, if combined into the same population, the individual genotype with the highest r would displace all others. That is, group differences in mean fitness were used as proxies for individual differences in genotypic fitness. It was acceptable to use group selection to understand individual selection in theory but it was not acceptable to use group selection to understand Nature.

The parameter r was considered *the* most important summary measure of fitness. Because r was maximized by individual selection according to the popular interpretation of Fisher's Fundamental Theorem, it was expected to be a nearly invariant, species-specific trait, lacking the additive genetic variance necessary for any further evolution. At the same time, however, types of selection that did not maximize r, such as frequency-dependent and density-dependent selection, were viewed as important to maintaining the high levels of genetic polymorphism that had recently been revealed by electrophoresis (Lewontin and Hubby 1966). However, neither frequency-dependent nor density-dependent selection maximized r because, with both types of selection, the fittest individual genotype tended to be the one that inflicted the most harm on its neighbors.

In short, I was taught a conspicuously dissonant view of group selection. It was a process useful in constructing theory but banned as an explanation of Nature. Maximizing population growth rate (r) was an axiom in some areas of research, but an axiom readily discarded in others, particularly those trying to explain genetic diversity. Most species lived in highly sub-divided populations, as a necessary condition for their persistence in the face of local extinctions, but none were believed to experience group selection. Extinctions and colonization could be frequent events in island biogeography, but, if so, they never generated group selection. Nearly everyone was using allozymes to quantify levels of polymorphism and calculating F_{ST}, the fraction of genetic variation among populations. Yet, F_{ST} was never discussed in relation to group selection, although the parameter had been designed by Wright for that purpose. Whereas large values of F_{ST} caused by drift were reported without comment, much smaller clinal trends (spatial increases or decreases) in gene frequency received considerable attention because they *might* be caused by individual selection. (Dave McCauley also recognized this ambiguity in the literature toward F_{ST}, and characterized it as "the Madonna of selection and the whore of drift.")

The Dawn of Kin Selection

As group selection fell from favor, kin selection was celebrated as a new explanatory paradigm in behavior evolution. Maynard Smith (1964) drew a distinction between group and kin selection based on "discontinuities in the breeding system." Discontinuities were essential for group selection but not for kin selection; kin selection could occur in randomly mating populations but group selection could not. Groups of kin could be virtual if individuals mixed randomly but behaved altruistically during encounters with close relatives but not when encountering more distant relatives. This required the further adaptation that individuals be able to recognize kin and modify their behavior accordingly. Everything from cuticular hydrocarbons in insects to chromosome numbers in wasps was interpreted as an adaptation for the purpose of recognizing kin. Flour beetles were a conspicuous exception, since larval cannibals could not recognize eggs of their own species let alone sibling eggs laid by their own parents. (Some of my colleagues were so convinced of the ubiquity of kin recognition that they believed that flour beetles did recognize kin but, for some other adaptive reason, ate them anyway.)

In behavior theory, there was another "little r," which governed the evolution of social behavior (Maynard Smith 1964; Hamilton 1964 a and b): genetic relatedness. High values of r promoted the evolution of social behavior; low values prohibited it. Through my statistics and probability courses, I recognized that r, the genetic correlation between individuals within groups, was equivalent to F_{ST}, the genetic variation among groups. In fact, the method for estimating heritability, using paternal half-sibs, was based on the fundamental relationship between the genetic differences among groups and the genetic correlation within them. The parameters, r and F_{ST}, were equivalent ways of looking at the same phenomenon, an opinion I expressed in a review of Dawkins' 1976 book, *The Selfish Gene* (Wade 1978c). My view was quickly labeled a "misunderstanding" (Dawkins 1979). In chapter 6, I discuss our experimental and theoretical studies designed to directly address Maynard Smith's definitional distinction between kin and group selection.

I thought there were additional problems with kin selection theory. It assigned each individual an *inclusive fitness* composed of two parts. The first was a direct effect on fitness based on the individual's own genotype as was standard practice in population genetic theory (cf. Table 2.1). The second was an indirect effect, the fitness benefit of the social behavior received from others and weighted by r. The weighting determined the identity of the genetic others and, therefore, what they could do for the individual's reproductive fitness. This fitness benefit from relatives was assumed to scale linearly with the number of kin altruists, so that members of a group with four altruists received twice the fitness benefit of those in another group with only two altruists. Both parts of inclusive fitness were interpreted as properties of the individual; it was assumed that their total was maximized by kin selection.

The consequences of social interactions were known to be highly sensitive to density and genetic context in simple laboratory organisms, but this was not taken into account in kin selection theory. For example, the fruit fly studies of Lewontin and his students (Lewontin 1955; Lewontin and Kojima 1960; Lewontin and Matsuo 1963) had shown that the genotype with the highest viability at one density did not enjoy the highest viability at another. Similarly, if housed with one genotype, the viability of a focal genotype might be high, but if housed with a different genotype, its viability might be considerably lower. The linearity assumption in inclusive fitness theory seemed ecologically unrealistic since, in all other theoretical treatments, interactions changed in strength with population density and gene frequency.

More problematic than the linear ecology of interactions was the maximization of fitness. Fitness was not maximized in the existing evolutionary theory of competitive interactions (Lewontin 1955; Wright 1969; Roughgarden 1971), where selection was frequency-dependent and density-dependent. These exceptions to fitness maximization were often highlighted because they were a source of dissonance between Nature and Fisher's Fundamental Theorem (e.g., Lewontin 1970). Thus, although kin selection was enthusiastically accepted by behaviorists, its reliance on fitness maximization, especially for traits sensitive to density and frequency-dependence, led it to be viewed skeptically by evolutionary geneticists. In addition, at Chicago, kin selection was suspect for its relatively uncritical application via sociobiology to the human social and political realms. It was seen by some as a modern resurgence of eugenics.

Most viewed kin selection as a type of individual selection—selection among individuals for adaptations to help (or at least not to harm) genetic relatives. Mayr (1988) later summarized this view: "kin selection is not group selection at all but rather a subdivision of individual selection." My training in probability and statistics implied just the opposite: kin selection was group selection where the groups consisted of genetic relatives. From my perspective, r was significant because it measured the among-group genetic variation. The greater the genetic variation among groups, the stronger was group selection.

Instructors and seminar speakers touted the example of the nine-banded armadillo. Because this species has litters of genetically identical quadruplets, siblings share the highest possible value of genetic relatedness ($r = 1$). Some Chicago faculty suggested to us that someone should devote their dissertation research to discovering kin selection and the hidden altruistic behavior of this species. Dawkins too (1976, p. 100–101) made a similar suggestion, "Nine-banded armadillos are born in a litter of identical quadruplets. As far as I know, no feats of heroic self-sacrifice have been reported for young armadillos, but it has been pointed out that some strong altruism is definitely to be expected, and it would be well worth somebody's while going to South America to have a look."

This made no sense at all to me: basic Darwinian logic prohibits individual selection of any kind from acting within armadillo families. Identical quadruplets are, by definition, not genetically variable. Within a single family, there is none of the heritable variation essential to individual selection. Thus, if there was selection for altruism toward litter mates in armadillos, it had to be based *entirely* on group selection *between* families (Wade 1978c).

I believed that kin selection offered an opportunity for conceptual arbitrage between statistics, evolutionary genetics, and behavior evolution, but it was not until almost a year later that group selection became the focus of my dissertation.

The Transition from Field to Laboratory Research

In the spring of my second year, I took Park's Field Ecology course, conducted in the vernal ponds of the Cook County Forest Preserves (Figure 3.1). These ponds had been discovered in the late 1920s by Park and his doctoral advisor, W. C. Allee, considered by many one of the pioneers of sociobiology. My course project was a study of adult mating behavior and offspring development in the eight species of amphibians found in the ponds. These data became the foundation of my first dissertation proposal designed to test the hypothesis that spring-breeding was timed to minimize predation on tadpoles yet permit metamorphosis before a pond dried up in early summer. My doctoral committee members were Drs. T. Park, M. Slatkin, R. Inger (curator at the Field Museum), and D. B. Mertz (of the University of Illinois at Chicago). I found that testing my hypothesis was very difficult because

FIGURE 3.1 The author as a graduate student pursuing tadpoles in a vernal pond in the Cook County Forest Preserves in April 1973.

there were frequent local extinctions of the amphibians and their predators. Some species not only went extinct between years but they also appeared in ponds where they had been absent the year before. The ponds were the embodiment of Levins' dynamic metapopulation but, as a practical matter, local extinction and colonization made experimental studies difficult.

About this same time, Park, as part of my training, asked me to review a manuscript for publication from the doctoral work of his former student Dr. M. R. Nathanson (1975). Nathanson had studied the effects of resource renewal on population growth rate in *Tribolium*. When I read his manuscript, I immediately realized how I could use the *Tribolium* system to experimentally study group selection. Nathanson had set up a very large number of populations, each with one or two species of flour beetles. For his experimental treatments, he had varied the rate of resource renewal, replenishing the flour every 30 days for some populations, but every 60, 90, 120, and up to 270 days in others. He also included a "never replenished" treatment, where all populations were certain to go extinct when the resource was exhausted.

Nathanson's study was timely because the rate of local resource renewal was an essential variable in the models of theoretical ecology. Although he was interested primarily in the effect of resource renewal on competitive outcome, Nathanson included single-species controls in his study just as Park had done in his classic experiments on competition. In his single-species populations, Nathanson found that extinctions became more frequent as the interval of resource renewal became longer. In fact, Nathanson discovered that he could *predict* the probability of a population's extinction: *populations with a higher rate of increase in the initial generations were more likely to go extinct, while the more slowly growing populations were more likely to persist.*

Park's *Tribolium* system had all the necessary features for empirically studying group selection! Where theory asserted that individual selection maximized *r*, Nathanson's data demonstrated that scarce resources favored reduced *r* by differential extinction, one of the mechanisms for group selection. In *Tribolium*, the important ecological trait *r* affected both individual fitness and population mean fitness. As a result, I could subject an array of *Tribolium* populations to group selection. Furthermore, I would not have to wait for the beetles to drive themselves extinct, which, in the laboratory, took many months or years. Instead, I could census populations 30 to 40 days after founding and predict, on the basis of the counts of adults, which populations would go extinct and which would persist. As the agent of group selection, I could use these predictions to decide which populations to proliferate and which to extinguish. In this way, I could impose 7 or 8 generations of artificial group selection within a year. Imitating artificial individual selection experiments, I proposed to "group select" in two directions: for increased r in one metapopulation and for decreased r in another. A third, control metapopulation without group selection was essential because I could not monitor individual selection within beetle populations.

The outcome of the experiment was clear, at least in theory: individual selection would maximize r and group selection would fail (Lewontin 1965; Charlesworth 1972). Some faculty held this view so strongly that they argued that my "control" metapopulation was unnecessary. However, if the group selection failed, a "control metapopulation" would allow me to test whether or not individual selection during my experiment was increasing or maximizing r. Park emphasized that, for the sake of my dissertation, I needed an experimental design where even a "failed" outcome would be of interest. This was especially important in my case because most expected my experiment to fail.

Prior to setting up the experiment, several concerns needed to be addressed. The central *genetic* issue was whether there was sufficient heritable variation for r in a *Tribolium* population. With natural selection relentlessly acting to maximize r, even in laboratory populations, theory said that no heritable variation in r should exist. Without heritable variation, artificial selection at any level would fail. However, Park's classic strains of both *T. castaneum* and *T. confusum* differed greatly in their population growth rates (Lloyd 1968). Hence, by mass mating equal numbers of males and females from these strains with very different values of r (Park et al. 1964), I believed that I could reconstitute the variable-r ancestral wild strain from which they had been derived. My synthetic strain would have the necessary heritable variation in r to begin my experiment. This mass mating was the origin of the *T. castaneum* c-SM stock (signifying "c-Strain Mix") that we continue to use in my laboratory to this day.

Although I had already defended a dissertation proposal based in the vernal ponds, Park encouraged me to draft a second proposal for a group selection experiment (Figure 3.4, see the proposal at the end of this chapter). Most believed that group selection to reduce r would be the key treatment in my experiment because it would oppose the maximization of r by individual selection. It would be *the* test case of William's principle of parsimony: opposing levels of selection. However, given the known density-dependence of life-history traits in *Tribolium*, I was not so sure that individual selection would maximize r. Park shared my concern and supported my imposing group selection for increased r in one treatment and decreased r in another.

One of these two group-selection treatments would have to be in the direction opposite to individual selection and thus be the "test case." Park and I agreed that a control metapopulation without group selection would reveal the direction of individual selection acting on r if I measured the trajectory of change in its mean population fitness over the course of the experiment.

My original proposal had only three treatments: group selection for increased r, group selection for decreased r, and a no–group selection control. However, during my proposal defense, D. B. Mertz (and, later M. B. Lloyd) argued that I should include a "random extinction" treatment as a "second control" for the extinction and recolonization process. The inclusion of this treatment turned out to be extremely important in the analysis and interpretation of my data. Also at my proposal defense, M. Slatkin persuaded me to use 16 adults whenever I founded a new population. With fewer beetles, he was concerned that inbreeding depression would compromise the experiment by reducing mean fitness. With more than 16 founding beetles, he worried that weak random drift (on the order of 3% $\{1/[2*16]\}$ per generation) would not create sufficient among-population genetic variation necessary for group selection, especially in the face of opposing individual selection. Thus, 16 founding adults was a compromise between enhancing drift and minimizing inbreeding depression.

The Committee and I also discussed whether I should use *Drosophila* instead of *Tribolium* for the experiment, since the generation time of the flies was half that of the beetles (14 days versus 28 days, respectively). With flies, I could have twice as many episodes of artificial group selection per year as I could with beetles. I argued that the live yeast, an essential component of the Chicago fly diet, were transferred to the larval growth medium on the feet of adult flies. This would make it impossible to determine the genetic basis of any observed change in r. If I saw a response to group selection, I would not be able to determine which species, fly or yeast, was responsible for it, since there did not seem to be an easy away to separate the two and measure the effects of one on the population growth rate of the other. Nathanson's postdoctoral research had recently demonstrated that healthy *Tribolium* cultures did not harbor any significant microbial fauna, so this was not a problem in the flour beetle system. I also argued that the *Tribolium* laboratory system was "more natural" in that this species had been commensal with humans for thousands of years. "Wild" beetle populations lived in small containers in pantries and granaries much as they did in Park's lab, while laboratory fruit flies were much poorer representations of Nature.

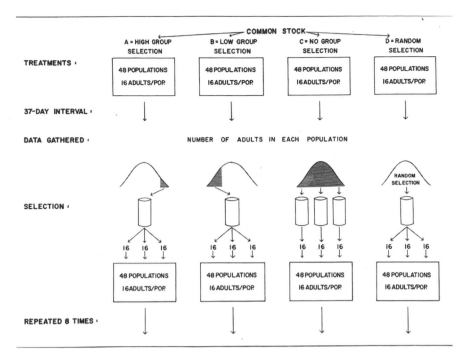

FIGURE 3.2 The experimental design depicted here was used in my doctoral study of group selection. The bell-shaped curves to the right of SELECTION represent the distribution of offspring numbers from the census data for each experimental treatment (x-axis: offspring number, y-axis: frequency of demes). The shaded area under each bell-shaped curve indicates the selected "parent demes" chosen by me, whose offspring were used to establish the next generation of metapopulation demes.

I already had an approved dissertation for field work and Park was adamant that I could not do thesis-level work on two projects at once. The rest of my committee believed that the *Tribolium* group-selection experiment, though nicely designed, would fail. Nevertheless, they found it promising enough to allow me to pursue it for five or six months, through the end of February, while the vernal ponds were frozen over and my field work was on hiatus. In February, we would meet again and decide, based on results, whether I should continue with the laboratory study of group selection or abandon it and return fulltime to pond work. With that caveat, I set up the experimental design depicted in Figure 3.2 (from Wade 1976, Fig. 1).

Results of the Group Selection Experiment

The response to artificial group selection was surprisingly rapid. After only two episodes of group selection, the treatments were statistically different in *r* and, after four generations, the differences between treatments were so large, that they were obvious during the census (Figure 3.3). Since population size in the control declined, the high–group selection treatment represented the canonical case at the heart of the group selection controversy: group and individual selection acting in opposing directions. Here, the effect of group selection was clearly evident. And, by the end of the experiment, populations from the high–group selection treatment averaged 100 beetles more *per population* than the controls, while populations from the low group-selection treatment averaged 60 beetles *per population* fewer than the controls. The high and low metapopulations differed from one another by a whopping 160 adult offspring per population! The differences were obvious without statistics.

In discussing my results with Levins, he speculated that I must be generating genetic variation for fitness among populations much more rapidly than anyone had previously thought possible. I realized that I could quan-

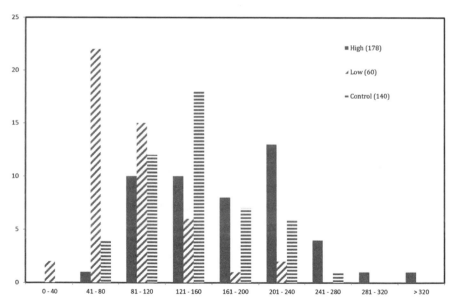

FIGURE 3.3 The frequency distribution of population sizes at generation 4 for the high– (solid gray), low–(slanted stripe), and control (horizontal stripe) group selection experiments. The average values are in parentheses.

titatively test Levin's intuition using my "random group-selection" treatment. At each generation, populations in this treatment could be grouped according to "parentage" in the preceding generation. Just as geneticists grouped individuals by parents to assess the heritability of individual traits, I could group populations by parentage and assess "group heritability." Because the parent populations were a random sample of their metapopulation, I could partition fitness or growth-rate variation into within- and among-population components using one-way analysis of variance. When I plotted these variance components over time (Wade 1976 Figure 2; Wade 1977 Table 2), they confirmed Levins' insight. The among-population variation increased steadily from 0% to more than 80% in seven generations. Random drift was converting within-population variation in r into heritable differences between populations in r, thereby allowing a large response to group selection.

By analogy with studies at the individual level, I introduced the term *group heritability*, and the symbol g^2, for the among-population fraction of fitness variation. Unlike the heritability of individual traits, h^2, which appeared to remain constant or nearly so across generations of artificial individual selection, g^2 was variable and responsive to random drift. Extinction and colonization of demes in the metapopulation created the g^2 necessary for group selection. As g^2 increased, a metapopulation became more favorable for group selection. This was a novel result: group heritability (g^2) was dynamically different from individual heritability (h^2), the static parameter that governed individual selection.

A second finding from the random-selection treatment was also surprising: population size began to increase after the 7th generation (Wade 1977, Figure 3), just as group heritability peaked at ~0.83. Although extinctions were random, my protocol resulted in the larger parent populations sending out more colonizing propagules than the smaller ones. This was group selection, not by differential extinction, but by differential colonization. When g^2 became very large at generation 7, the weak group-selection arising from my unrecognized differential colonization was sufficient to cause a measurable response. (If I had instead taken the same number of propagules from every surviving deme, there would have been *no* group selection in this treatment as I had intended. We eliminated this type of group selection in later experiments [cf. chapter 4].) Coincident with the response to group selection, g^2 declined: group selection eroded the heritable variation for fitness at the population level just as individual selection eroded heritable variation for fitness at the individual level. Although my analyses and interpretation of the random-selection treatment were unplanned and post hoc, they allowed me to demonstrate group selection by two different mechanisms: (1) by differential extinction and dispersion in the high and low treatments, and (2) by differential colonization alone in the random extinction treatment.

The Results Pose Further Questions

My results raised as many questions as they answered. Population growth rate, r, was a composite parameter, influenced by many, many different traits. Which traits caused populations that started with the same numbers of adults to have such different values of r? Could I show that group and individual selection had been in conflict for one or more of the composite traits? Had the controls declined in average population size owing to individual selection for cannibalism or was some of the decline owing to inbreeding depression? How small was the "effective" population size of a group of 16 founding adult beetles?

In addition to these, there were many more general questions that could be answered only with further experiments. Would individual selection reverse and eliminate the group-selected gains if I were to relax group selection? Would a larger founding colony size, greater than 16 adults, diminish the effectiveness of group selection? Would migration between populations have prevented any response to group selection at all? Was there some unique aspect of *Tribolium* biology that allowed group selection to work in the laboratory but limited the implications of my findings for natural populations? The questions were not as surprising as how my results were interpreted by others. My colleagues believed that the questions would all be answered "Yes," so that my results, while unexpected, could be dismissed without further investigation.

Fortunately, the rapid response in the main experiment allowed me time to conduct what Park called an "ancillary" experiment as part of my dissertation. The question of most concern to me was the effect of migration on the response to group selection. On theoretical grounds, populations exchanging as few as one migrant every other generation ($\frac{1}{2}$ a migrant per generation) were expected to have equal gene frequencies at equilibrium in the absence of selection (Maruyama 1970; Crow and Kimura 1970). This was interpreted (incorrectly) to mean that a little bit of migration would reduce g^2 to 0. Migration was antithetical to g^2 and I had no migration at all between populations in my main experiment. From my fieldwork, I was certain that, when an organism's persistence in the face of local extinctions depended upon colonization, there would also be significant levels of migration. I was afraid that, in a metapopulation with migration, group selection would fail.

The Ancillary Experiment: Group Selection with Migration

To test the effect of migration on group selection, I used beetles from the "leftover" populations of the main experiment (generation 4) to found new metapopulations. The leftover populations were those from my experimental metapopulations that I would ordinarily have discarded, because their r was too small for the high treatment and too large for the low treatment. From the random-extinction treatment, I took leftover populations at random. From six of the control populations, I took all the leftover beetles. With these, I created four new stock populations, each with approximately 1,000 beetles, by mixing adults from four leftover high populations, seven random-extinction populations, twelve low populations, and six control populations. With offspring from these four stocks, I established 12 new metapopulations, each consisting of 18 populations founded with 16 adult beetles. (There were 3 metapopulations from each of the 4 stocks, for a total of 216 populations. The ancillary experiment was larger than the main experiment [192 populations]!)

On these metapopulations, I imposed three levels of migration. The first was 0% migration, which was essentially identical to the first experiment, allowing its replication. The second was 25% migration between pairs of populations. For example, in the high treatment, the highest [a] and the second highest [b] populations founded two new populations, one with [12a and 4b] adults and one with [4a and 12b] adults. In the control and random treatments, populations were paired randomly and exchanged comparable numbers of migrants. The third was 50% migration between pairs of populations, so it was similar to the second strategy, only the numbers of adults were equal [8a and 8b]. This protocol gave me four metapopulations (high, low, control, and random) for each level of three levels of migration (0%, 25%, and 50%).

Although there were many other possible ways to impose migration, Park and I decided that this type of mixing was a good first step, intermediate between the extremes of completely random or "island-model" mixing and the so-called stepping-stone models with mixing primarily between neighboring populations. Even so, I was not sure how migration would affect the response to group selection. Migration must reduce the effect of random drift on the variation among populations. The issue was this: To what degree would migration reduce g^2?

To my surprise and delight, the results of this second experiment showed convincingly that migration did not prevent a response to group selection. Group selection produced a significant response in every meta-population on which it was imposed. Although the high–group selection response was 2.5-fold greater without migration than it was with migration, I saw a significant response even for the highest level of migration. In the low–group selection treatments, migration had no discernible effect: the response to group selection was large and independent of the amount of migration. The data showed that migration was not antithetical to group selection as theory predicted!

These results also *replicated* the findings from the main experiment. Combining results from both experiments, in all cases, two with migration and two without, group selection successfully increased population size relative to a control with individual selection. All four cases of group selection for *decreased* population size were also successful.

Interactions Caused the Differences in Population Growth Rates

To complete my dissertation, I had to find out which traits had been genetically altered by group selection to cause the differences in r. Fortunately, population growth rate in *Tribolium* had been well studied by Park. It was known to consist of a series of component traits, which Park and his colleagues had divided into two categories: "primary characteristics" and "group interactions." Primary characteristics were measured on isolated single beetles or pairs of beetles. They included fecundity, egg fertility and viability (hatchability), larval survival to adulthood, development time, sex ratio, and body size. Group interactions were traits, like cannibalism and sensitivity of development to crowding, which could not be measured on isolated beetles.

Because the control mean had declined, I expected that cannibalism, a group interaction, would be key to understanding the metapopulation differences in r. First, in theory, cannibalism was always favored by individual selection because it increased the fitness of the cannibal while decreasing the fitness of its victim. Secondly, cannibalism had been shown to have explanatory power in the analysis of Park's ecological studies of competition: the more cannibalistic strains were often better competitors. These expectations propelled me through a series of exceedingly tedious, highly replicated, descriptive experiments.

The Primary Characteristics

I used the tried and true methods that Park and his students had developed for measuring individual traits in flour beetles, applying them to several populations from each of the four metapopulations of the main experiment as well as to the c-SM stock. In each case, I assayed the six primary characteristics mentioned above (Wade 1976, 1979a), starting with the earliest life stage, the numbers of freshly laid eggs. As expected, I found that females from the high-group-selected populations laid more eggs (24% more) than females from the controls. Unfortunately, so did females from the low-group-selected populations and from the random-extinction populations (22% and 24% more eggs, respectively). Females from the control populations and the c-SM stock laid identical numbers of eggs. Fecundity had increased in all treatments relative to the controls, but made explaining the large differences in population size more difficult. How could beetles from the low-lines lay so many more eggs, yet produce considerably smaller populations?

I found that egg hatchability was the same for all treatments and larval-to-adult viability showed only modest differences: ~10% more larvae survived to adulthood from the high populations than survived in the controls or in the ancestral c-SM stock. There were no significant differences among treatments for body size or sex ratio. Despite thousands of careful measurements, none of the first five primary traits differed among treatments in a way that could account for the large differences in population growth rates.

My last hope was development rate. In theory, the effect of shortening or delaying development on r was much greater than increasing or decreasing the number of eggs, the viability, or the longevity (Cole 1954; Lewontin 1965). In fact, reducing the duration of development by a single day, in theory, was equivalent to doubling the total number of eggs laid (Cole 1954; Lewontin 1965). For this reason, I was happy to record developmental transitions, from egg to adult, from one instar to the next, etc., every six hours, day and night, for 35 days for hundreds of beetles from several populations from each treatment and the c-SM stock. But, I did not find an explanation for the treatment differences here either. Development from larva to adult was faster in the high–group selection populations than in the controls, but the difference in rate of development was less than one day. Males developed significantly faster, not slower, in the low–group selection populations, although again, the magnitude of the difference was less than one day.

For all that effort, the primary characteristics revealed disappointingly little about the response to group selection. None of the small differences in any trait, alone or in multiplicative combination, explained why the high–group selection populations averaged 100 beetles more per population than the controls; why the controls declined conspicuously in size over time; or why populations from the low–group selected metapopulation averaged 60 beetles per population fewer than the controls.

Further confounding the interpretive power of these observations, I found large and significant differences among populations from the *same treatment* for each trait. For example, in one low–group selected population, egg numbers were 45% lower than those of control females, in another low–group selected population, they were 46% higher, and in yet another, they were equal! I was forced to conclude that, not only is population size a complex trait, but also there are many different ways that a group of beetles could achieve high, low, or intermediate population size. And, despite a strong response to group selection, there still remained a great deal of among-population variation (i.e., g^2) for many of the traits believed to be important components of lifetime fitness.

The Group Interactions

It was more illuminating to measure the group interactions, like the four cannibalism pathways (adults eating eggs, larvae eating eggs, larvae eating pupae, and adults eating pupae). Since adults both lay and eat eggs, I had to use a mark-recapture method devised by Park to assay adult-on-egg cannibalism. It required that I distribute 26,000 red-dyed eggs uniformly in the flour of small vials before adding uniformly aged adult beetles as potential cannibals. Later, I had to collect and count both the marked (red) and unmarked (white, newly laid) eggs. This huge effort was worthwhile, because it explained part of the response to group selection.

Adults from both the high– and low–group selected populations exhibited significantly different egg cannibalism rates than adults from either the control or the ancestral c-SM stock. Net fecundity (i.e., eggs laid minus eggs cannibalized) was in the right order: highest in the high populations, intermediate in the controls and lowest in the low populations (Wade 1979a, Table 6). Clearly, the "primary fecundity" of single females (see above and

Wade 1976) was not representative of the net fecundity for groups of inter-acting females.

Unfortunately, the other cannibalism pathways, though equally tedious to measure, were much less informative. Larval predation on eggs or upon pupae did not vary among treatments. Control adults consumed more pupae than adults from either the high–group selected demes or the c-SM popula-tion. However, adults from the low–group selected populations consumed on average 38% fewer pupae than the control adults—a highly significant difference, but in the "wrong" direction. Adult-on-pupal cannibalism, like single-female fecundity, also varied significantly among populations from the same source.

Conclusions about Causes

The primary traits did not explain the differences in population growth rates in my experiments like they did for the Park strains (Park 1954). Some group interactions, like adult-egg cannibalism, showed clear differences in the right direction. I could infer that, at least for some interaction traits, group and individual selection had acted as opposing forces. Part of the dif-ficulty in accounting for the large treatment differences was the surprising heterogeneity of trait means among populations from the same treatment. Ultimately, none of the traits, alone or in combination, came close to ac-counting for the large treatment differences in average population size. I was forced to "invoke as yet unstudied interactions between traits or be-tween certain traits and population density to explain the treatment differ-ences" (Wade 1979a, p. 764). Two years later, my first postdoc, D. McCauley, designed an experiment that revealed how interactions between beetles within populations caused the large differences in population growth rates.

The responses of my colleagues to these data were interesting. Many were disappointed to learn that their favorite trait did not explain all the differences and drew the conclusion that *Tribolium* biology must therefore be strange or unusual. Others reasoned that, once a trait connected to indi-viduals was shown to have any explanatory power, then my experiment had not been real group selection after all, but rather some sort of cryptic indi-vidual selection. This viewpoint was particularly difficult and frustrating to counter. It suggested that, if I discovered the mechanism behind my group-selection results in the traits of individuals, then the process by which I had produced the results was somehow negated.

It is my intention to design and carry out an experiment
to answer the question,"does group selection occur?". The nature *can; does*
of the question places certain logical constraints upon the
experimental design. It iS first of all necessary that the design meet the three principles embodying the process of evolution by natural selection(Lewontin, 1970):

1. the units involved must differ from one another in
some biological respect. (The groups of Tribolium
castaneum will, most probably, differ in standing
population size at day 30.)

2. these differences between units must result in differential rates of survival. (This condition will
be imposed upon the groups by the experimenter, i.e.
all but those four replicates with the smallest standing populations at day 30 will be discontinued.)

3. those differences which contribute to the differential survival *reproduction* of the units must in some way be heritable. (Five propagules of eight individuals each
will be selected at random from the four minimum populations and these propagules will become the next
focus of selection.)

If group selection exists, it should act regardless of
whether individual selection is acting in the same direction,
in the opposite direction, or not at all with respect to the
trait in question. However, in order to obtain a definitive
answer to the experimental question it is necessary to adopt the
worst case. The second constraint is therefore that individual

selection must be acting in the opposite direction with respect
to the trait in question.

Thirdly, there must be some criteria by which it can be
determined, at the end of the selection regime, whether or not
a significant change has occurred in the population.

The specific design that I would like to discuss sprang
from a statement by Dr. Michael Nathanson to the effect that
standing population size at day 30 is a good predictor of the
relative time to extinction of a population of Tribolium castaneum when on an infrequently(150 days) renewed food resource.
Low population numbers at day 30 indicate longer survival times
and this holds whether Tribolium castaneum is husbanded alone
or in competition with Tribolium confusum. Thus, I would like
to select for smaller populations at day 30 and after two years
of selection place propagules from the final populations on a
schedule of infrequent flour renewal, both alone and in competition, to determine whether or not a change has occurred. Group
selection would be acting to decrease "r", the intrinsic rate of
increase.

In order to assure that individual selection is acting in
the opposite direction, i.e. to increase "r", I have considered
both of the following formats:

A) establishing populations of 8 adults(sex ratio of
unity) on 8grams of standard medium at 29°C and 70per cent humi-
ity. Assay populations at day thirty and make a decision to
select those four populations with the lowest standing crops
(adults and larvae). Allow the populations to remain intact for
an additional 30 to 60 days in order that individual selection
will act in the typical way to increase "r", i.e. those individ-

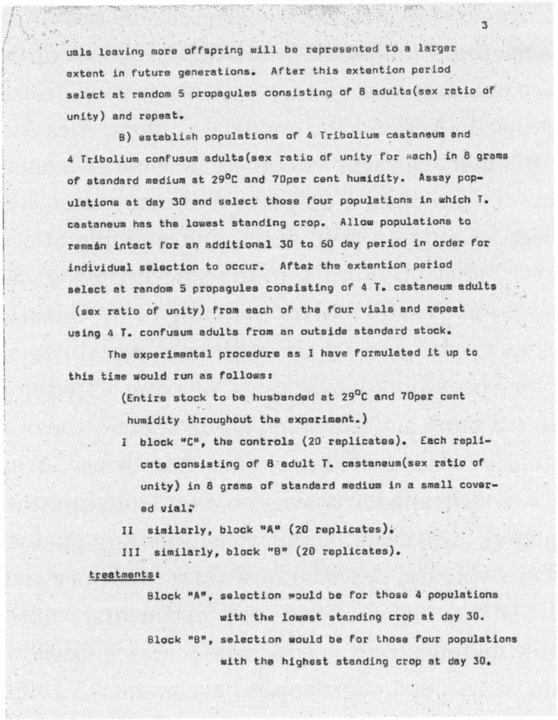

uals leaving more offspring will be represented to a larger
extent in future generations. After this extention period
select at random 5 propagules consisting of 8 adults(sex ratio of
unity) and repeat.

    B) establish populations of 4 Tribolium castaneum and
4 Tribolium confusum adults(sex ratio of unity for each) in 8 grams
of standard medium at 29°C and 70per cent humidity. Assay populations at day 30 and select those four populations in which T.
castaneum has the lowest standing crop. Allow populations to
remain intact for an additional 30 to 60 day period in order for
individual selection to occur. After the extention period
select at random 5 propagules consisting of 4 T. castaneum adults
(sex ratio of unity) from each of the four vials and repeat
using 4 T. confusum adults from an outside standard stock.

    The experimental procedure as I have formulated it up to
this time would run as follows:

    (Entire stock to be husbanded at 29°C and 70per cent
    humidity throughout the experiment.)

    I block "C", the controls (20 replicates). Each repli-
    cate consisting of 8 adult T. castaneum(sex ratio of
    unity) in 8 grams of standard medium in a small cover-
    ed vial.

    II similarly, block "A" (20 replicates).

    III similarly, block "B" (20 replicates).

treatments:

    Block "A", selection would be for those 4 populations
        with the lowest standing crop at day 30.

    Block "B", selection would be for those four populations
        with the highest standing crop at day 30.

FIGURE 3.4 The doctoral proposal for an experimental study of group selection that
I defended before my committee, consisting of Drs. T. Park, M. Slatkin, R. Inger, M. B. Lloyd,
and D. B. Mertz. It is difficult to imagine distributing a proposal this lacking in scholarship and
detail to a dissertation committee today.

Writing and Defending My Dissertation

At spring break in 1975, M. Slatkin accepted an invitation to visit Japan for
several months, beginning in June. I had to write, submit, and defend my
thesis before he left, if I wanted to complete my PhD before the fall. And, I
had to do it right in the middle of the relentless six-hour development assays. The pressure was on, since my future position at Chicago as an assistant professor was conditional on my completing my PhD. I wrote chapters
out in longhand and handed them in to Park, who, overnight, would heavily
edit the first two or three pages in his customary purple ink. He expected me
to use his detailed commentary on the first few pages as a model for editing
the remainder. Each morning in his office, we would discuss my previous
day's writing effort in terms of both content and style, sentence by sentence.
I would then type a revision and hand it to Slatkin for additional critique.

    Slatkin was concerned that the first draft of my dissertation was "too experimental" for a student in the Department of Theoretical Biology. In response, I developed a computer simulation of the group-selection process that incorporated, at least to a first approximation, features of my experimental protocol. (A friend and medical student, Alex Aisen, who was working on computer analyses of magnetic resonance imaging, wrote the actual simulation.) The simulation permitted me to alter the strengths of individual selection, random drift, extinction, and migration. It produced results that were sufficiently similar to the experimental data that Park endorsed the exercise as a useful tool for planning future research.

I was anxious to plan and set up additional experiments, varying the number of colonists and migration rates, relaxing selection and tracking the loss of selected gains, and imposing group selection on other traits, like competitive ability. I also wanted to experimentally test Wright's Shifting Balance Process, which relied so heavily on interdemic selection. In the next chapter (chapter 4), I will describe my transition from graduate student to faculty member. I will also describe how Dave McCauley and I, with NIH funding (with Slatkin), began a series of experimental studies that solidified many of the findings from my doctoral research. Together, we moved my research in directions I never could have predicted at the time of my thesis defense.

4 Career Beginnings and Science after the Thesis

Career Options

Park (Figure 4.1) retired in May 1974, just as I defended and completed my PhD. He was pleased that the *Tribolium* laboratory system that he had spent his career developing would continue in my hands at Chicago, albeit with a more genetic focus.

The four classic "Park strains" of *T. castaneum* and *T. confusum*, used in so many ecological studies of competition and demography, had been developed by his student, M. B. Lloyd, using only two episodes of family selection, separated by 5 generations of no selection at all. These strains varied several-fold in equilibrium population size, competitive ability, cannibalism rates, and several other life-history traits. And, the differences among them had persisted stably for decades without further selection. For this reason, Park was convinced that group selection could be a highly effective force for changing ecologically important traits like competitive ability and he encouraged me to explore it. He and his colleague, Dr. Sewall Wright, had long speculated about using the power of the *Tribolium* system for investigating genetic evolution in a realistic ecological context. Such experiments would combine not only theory and experiment, but also the fields of ecology and evolutionary genetics. Park considered my research to be a significant step in that direction

FIGURE 4.1 Dr. Thomas Park in our home at 1700 East 56th Street in June 1974.

and arranged for me to meet Wright and discuss my future research plans (see chapter 9).

In the early years of my career, I would not have attempted so much experimental work without Park's support and his willingness to discuss my experimental designs at great length. These discussions took place in his home in the early morning over coffee or in the late afternoon and into the evening over potent gin and tonics.

My career options, before the group selection experiment had produced results, appeared to be very limited. Family commitments constrained me to working in Chicago for at least the next two years and there were very few faculty positions at all nationwide in population biology. Because my doctoral experiments were still in progress, I had no publications and no thesis. After graduation, I thought I could capitalize on my Boston experience and be a substitute high school teacher in the Chicago Public Schools. Fortunately, two assistant professorships were advertised within my geographic reach, one at the University of Chicago (U of C) and the other at the University of Illinois at Chicago (UIC). Although U of C had policies against departments hiring their own students, my PhD was in the Department of Theoretical Biology and the position was in Biology, a small but administratively crucial distinction.

I applied for both positions, but with no publication record, Park emphasized that my success or failure would depend much more heavily than most on my letters of recommendation. These letters posed their own problem. Although the members of my doctoral committee knew my work well, their evaluations were compromised because they were faculty at the universities doing the hiring. Letters from faculty at other universities would be better, but few were aware of my unpublished experiments. Thus, to get the necessary outside letters, Park and Slatkin arranged for me to give seminars at Harvard and Northwestern. As a lowly graduate student, I found it excruciatingly difficult to travel to a university, meet with faculty all day, and, after presenting a seminar on a controversial topic, ask my host for a letter of recommendation. I was not optimistic before or after either trip.

Nevertheless, I was invited to interview for both positions and presented my job seminars in the fall of 1974. I suspected that both interviews were courtesy interviews arranged through collegial pressure on the respective search committees by my doctoral committee members, rather than invitations based on my accomplishments. The U of C interview was exceptionally draining. Many faculty I met with during the interview had either had me as a student in class or were in other biological disciplines and somewhat envious of the renown surrounding the ecology and evolution group. Many were openly skeptical of my viability as a candidate. However, unknown to me, I had already made a positive impression on a handful of faculty during my first experience at graduate teaching. After Lewontin left U of C for Harvard in 1972, Slatkin had taken over the graduate Population Biology course, which was almost entirely theoretical in content. While Slatkin was away giving seminars, he asked me to teach two weeks of his class, covering the topic of B. Charlesworth's life-history theory. The former Biology Department chair, Dr. W. Baker, attended all four of my lectures. Baker had considerable influence within the department and Park informed me years later that I had had Baker's support prior to my application for assistant professor on the strength of those graduate lectures.

After the interviews, I heard nothing for months and months, except comments about the merits of the other interviewees. I learned that U of C was offering its position to Dr. Ross Kiester, a Harvard-trained tropical field ecologist with formidable mathematical abilities, when he arrived on campus for a second visit. I assumed that that was it for me as far as U of C was concerned. When Dr. E. Spiess, the UIC chair, sent me an offer letter, I was prepared to accept it on the spot without negotiation. At this point, Park intervened and counseled me, without explanation, to hold off answering Spiess' letter for one week. At the end of that week, on a Friday afternoon as I was counting beetles, the Biology Department chair, Dr. Hewson Swift, came into Park's lab and asked whether I would accept a position as assistant professor at U of C if it were offered. Somewhat stunned, I said yes. He then asked if I would accept a salary of $16,500 per year. It was an enormous sum for someone who had lived for four years on an annual graduate stipend of $2,600. I said yes again and Swift replied, "In that case, I will have a written offer for you by Monday." The offer arrived in my mailbox Monday as promised and included $750 in set-up funds and an assurance that Park's lab would be repainted. I accepted the offer immediately. (I blew more than all of my set-up money on a hand-held, programmable Hewlett-Packard HP-65 calculator. It cost $795 and was advertised as "the smallest programmable computer ever!")

I was both thrilled and apprehensive to have a position at Chicago. On the positive side, I could keep my experimental work going without interruption in Park's lab. Slatkin and I had secured NIH funding, so I could employ Park's dedicated technician, Mrs. Ora Lee Watts, who had more than 17 years of experience with *Tribolium* husbandry. And, although I may have been the lowest paid faculty member in the entire Biological Sciences Division, I was still a faculty member at the University of Chicago!

On the negative side, I felt an acute sense of responsibility as a new member of an intimidating academic tradition that included F. R. Lillie, W. C. Allee, H. C. Cowles, A. E. Emerson, S. Wright, T. Park, G. W. Beadle, R. C. Lewontin, and R. Levins. Throughout my time at Chicago, their portraits hung in the Lillie Room of the Zoology building near my lab. Park had an encyclopedic grasp of University and Department history, both intellectual and administrative, and he expected me to know these men and their work.

FIGURE 4.2 It was difficult to imagine myself as a beginning assistant professor at the University of Chicago in late summer 1975. The transition from graduate student (Fig. 3.1) to faculty member included "making myself more presentable."

He expected me to conserve the institutional memory of the Department's intellectual tradition as well as measure up to it. That was real pressure.

An immediate and practical problem for me was timing. My assistant professorship started October 1, 1975, but my training grant support ended in August. This meant I would have no income for the month of September but still faced rent and other bills. Fortunately, someone at the Chicago hospitals was testing experimental asthma drugs. He paid participants $325 a study and I volunteered, since I had asthma. It was very good money and, with the guidance of the other longer-term volunteers in the study, I was accepted into a second study. By taking experimental drugs, I made $650 in a single month—three times the training grant rate!

Group Adaptations and Group Selection

My first priority upon graduation was publishing but my empirical work did not fit well within the existing literature. In the literature, the group selection controversy was framed as a "who benefits" question. For example, Cassidy (1978, p. 582) stated that *"the possessor and beneficiary of a trait"* defines the level of selection. My focus on group selection as a process rather than on group adaptation as an outcome made it different from most of the published work (Wade 1977, 1978a). I found my work relegated to the periphery of the controversy it was designed to address. In the ensuing decade, a process-focused view on the levels of selection would develop in history and philosophy of science (Brandon 1982; Michod 1982; Sober 1984; Lloyd 1988; Hull 1980, 1988). Just as problematic was our notion of group heritability, g^2. It was a new concept in the process-centered definition of group selection that, while key to interpreting my results, was irrelevant to most topics in evolutionary biology, including adaptation-oriented debates about group selection.

Williams' popular parsimony argument demanded the a priori "privileging" of individual selection over group selection as an explanation for adaptation. This was not at all unusual at that time; individual selection was similarly privileged in relation to all of the other evolutionary forces—that is, mutation, migration, and random genetic drift. However, I was convinced that the present value of a trait did not contain, by itself, enough information to reconstruct its past evolutionary trajectory or the processes that had shaped it. Awarding an adaptation to a lower-level process by parsimony overestimated the strength of individual selection whenever the two levels acted in the same direction as well as when the higher level of selection was acting alone. If both selection levels were acting, but in opposite directions, parsimony underestimated the strength of both.

There was an additional source of bias: several of the assumptions of the mathematical models of group selection biased the subsequent analysis and conclusions in favor of individual selection (Wade 1978b; Lloyd 1988; see below under Reductionist Biases). For all of these reasons, I continued to focus my research away from group-level adaptations après Wynne-

Edwards, Williams, Dawkins, Sober, or D. S. Wilson and toward group selection as an evolutionary genetic process. But, if the group selection debate was not a good context for my results, what was?

Animal Breeding and Family Selection

There were no other published experimental studies of group selection at the time to use as models. However, there were many examples of family selection in the literature on the domestication of plants and animals. That literature contained extensive discussions of the conditions under which family selection was more effective than individual selection (see chapter 2). Furthermore, those discussions centered on individual traits of low heritability, such as my trait of population growth. This research tradition offered a robust conceptual and experimental context for the results of my studies.

I found selection for increased milk yield in cattle a particularly interesting example of the difference between the process-centered and the trait-centered viewpoints. In dairy breeding practice, cows were selected on the basis of individual milk yield, but bulls were chosen by family selection, based on the average milk yield of their sisters or other female relatives. This was necessary because bulls do not give milk but do pass genes affecting milk yield to their daughters. For dairy cattle breeders, the combination of individual selection on cows and family selection on bulls was more efficient than individual selection acting alone on only the females. Selection was focused on a single trait, but the level of artificial selection differed between the sexes.

This particular example of sex-differences in the selection process made me realize that there was an interesting issue regarding sexual selection for exaggerated male traits, an emerging topic of active research (Ghiselin 1974; Williams 1975; Dawkins 1975; and later, Maynard Smith 1978). Sexual selection was considered a strong evolutionary force for two reasons. First, in sexually selected traits, males of closely related species were more different from one another than were females, a phylogenetic signal of rapid evolution. Second, R. A. Fisher had described the process of "run-away" sexual selection, whereby female mating preferences became exaggerated along with the male traits they preferred. This was a coevolutionary explanation involving interactions between male and female traits. However, I wondered how selection for exaggerated traits in one sex, but opposed by selection in the other sex, could be so strong. A breeder favoring the best cows for milk yield but the worst bulls would not enjoy much success at increasing milk production. In my later work (Wade 1979b, 1985; Shuster and Wade 2003), I took up this issue of sexual selection for exaggerated male traits, but the idea was stimulated by my interest in family selection on a female trait, milk yield.

The literature on artificial selection also informed our future experiments in multilevel selection. That literature gave me confidence that, despite the overt anti–group selection bias within the field of evolutionary biology, there was an important genetic question that could be addressed empirically with our *Tribolium* model. It is possible that our adoption of concepts and language from the literature on animal and plant domestication drew the attention of animal breeders to our work (e.g., Muir 1996) where it is now an important tool for increasing yield and enhancing animal welfare, particularly in laying hens (Wade et al. 2010; Cheng and Dennis 2011).

Kin Selection as an Alternative to Group Selection

By the completion of my doctoral degree, kin selection theory had taken center stage as an attractive alternative to group selection. Williams (1966), Dawkins (1976, 1982), and others argued that the gene, an entity within an individual, was always the ultimate beneficiary of the adaptive process (e.g., Hamilton 1964; Dawkins 1978). Because a gene benefits, they concluded that selection acted at the level of the gene, below the level of the individual. This "gene's eye view" of evolution was irresistible to many, especially those concerned with the evolution of social behavior, although it has many conceptual disadvantages (Lloyd 1988, p. 117).

The primary disadvantage of the "gene's eye view" of evolutionary genetics is that it lumps all models, of all levels of selection, into one. In theoretical population genetics, all models of adaptive evolution involve positive change in gene frequency. In every case, the marginal fitness of a favored allele (the genic fitness) must exceed one, the mean of *relative* fitness of the population, in order for the frequency of a gene to increase by selection and spread through a population. Although every selection model has an expression for marginal genic fitness, modeling change in gene frequency is not a sufficient depiction of the evolutionary process for complex, multigene models (Lloyd 1988). Furthermore, marginal genic fitness is an abstraction and not a "property" of a gene, although Williams (1966, p. 57) asserted otherwise. Lastly, it was well established that marginal genic fitness did not map uniquely onto a single selection process and therefore was not useful for distinguishing among different kinds of selection (Lewontin 1974; also reviewed in Sober and Lewontin 1982). Change in gene frequency is the outcome of selection but it is not a depiction of the process of selection. In my experiments, some genes enjoyed a net marginal benefit, not as an extension of individual selection or by reduction to genic selection, but rather as a result of the group selection that I imposed upon an experimental metapopulation.

Kin selection and inclusive fitness theory combined the gene's eye logic with the central tenet of Mendelian genetics — namely, an individual and its biological relatives share similar genes. An individual practicing altruism by helping its relatives is benefiting copies of the same gene in those relatives, at least on average. For this reason, it was argued that the concept of individual fitness had to include one's relatives, an extension of a gene's fitness domain called "inclusive fitness." If an individual helped its relatives sufficiently more than it hurt itself, altruism would evolve by what some called "individualistic kin selection." Because the average fitness benefit is greater for genetically close relatives than it is for more distant relatives, a parameter quantifying degree of genetic relatedness, r, became central to the theory. It was a simple conversion factor relating an individual's fitness cost to its relatives' fitness benefit. The accepted conclusion was that "even in the most specialized 'supraorganismic' social insects, selection was still operating at the individual level" (West-Eberhard 1975, p. 15). Indeed, the so-called Central Theorem of Sociobiology at the time of my doctoral work was that all organisms will behave as though they are maximizing their inclusive fitness (Barash 1977).

In contrast, in the statistical genetics literature of animal breeding, relatedness, r, measured the genetic variation *among families* and determined the efficacy of family selection. As r increased, the evolutionary process changed from a more or less even mixture of within-family and between-family selection, when r was ½, toward pure family selection when r equaled 1 (as in the case of the armadillo families). In the classic, paternal half-sib design of animal breeders, paternal siblings from different mothers share a genetic correlation of 0.25 owing to the common sire. It is for this reason that the among-sire variance estimates ¼ of the additive variance, the numerator of h^2. Thus, to the animal breeder, kin selection was a type of family or group selection. In some sense, a dairy cow's milk yield could be considered "altruism" toward her brother bulls, since these males got to live or at least to breed, not on their own merit, but on that of their sisters. (Some would say it is weak altruism, at best, however, because cows with high milk yield are themselves favored by breeders, so they are not really sacrificing anything in "promoting" their brother's fitness.)

My interpretation was that kin selection was an example of multilevel selection, different from the version of individual selection underlying sociobiology (E. O. Wilson 1975).

Family Selection

To test my understanding of the equivalence between family selection and kin selection, I attempted a population genetic model for the evolution of altruism by family selection (Wade 1979c); it was my first venture into mathematical modeling. I discovered that G. C. Williams twenty years earlier had coauthored a model of family selection for social behaviors, "Natural selection of individually harmful social adaptations among sibs with reference to social insects" (Williams and Williams 1957). The authors specifically mentioned Wright's (1945) earlier work on this problem using group selection, saying (p. 32), "In this paper, we will also use favorable between-group selection to balance unfavorable within-group selection, but our groups are sibships, not Mendelian populations." They explicitly described multilevel selection with individual selection against a "social donor" gene and between-family selection favoring it (p. 34):

> Competition within the sibship will always place the donors at a disadvantage. Competition between sibships, however, is also a factor in the survival of the donor gene. Sibships that contain donors will be favored if A [social benefit to sibs] is greater than D [cost in fitness to the individual]. Selection, then, would be the resultant of two opposing forces, adverse selection within sibships and favorable selection between sibships.

They concluded with a parsimony argument that G. C. Williams would later use so effectively against group selection: "Until such evidence is forthcoming, it is unnecessary to postulate selection at a higher level than the family to explain the development of insect societies" (p. 38). Unfortunately, in their analysis of the problem of social evolution, they used a fitness ratio approach pioneered by Haldane, rather than gene frequency change; this modeling choice greatly complicated their analysis. They ended by stating that they had "not been able to derive equations that make it possible to determine the net effects of such selection on gene frequencies" (p. 35). As a result, they did not discover the central role of genetic relatedness in the evolution of individually disadvantageous social behaviors or the formal relationship between individual selection, family selection, and Hamilton's Rule (Wade 1979c, 1980a).

In my work, g^2 was analogous to r, but applied to whole populations instead of families. In my doctoral experiment, by generation four, g^2 in the *Tribolium* metapopulations had exceeded 0.25, the genetic variance among half-sib families; by generation 6, it was greater than 0.50, the value of r for full-sib families. And, it continued climbing until it reached a peak value of 0.80. If selection among families was a viable process for much smaller values of r, it was no wonder that group selection in *Tribolium* had been effective given such high values of g^2.

However, I needed to demonstrate that my *Tribolium* populations were much larger evolutionary entities than families. It was possible that 16 breeding adult beetles might be functionally equivalent to a single family. This could happen if the variation in family size within a population was so large that, on average, only one male and one female out of the founding 16 produced all or almost all of the offspring. If so, it would reduce my group selection to "just" another case of family selection.

Reductionist Biases

Not only was the "gene's eye view" of evolution becoming popular, but also there were other biases acting against group selection. These took the form of a series of assumptions used routinely to simplify mathematical models and their analysis (Wade 1978a), which, together, constituted what Dr. W. Wimsatt (2007) calls a *reductionist bias*. The assumptions included (1) island-model migration, where thorough mixing every generation eliminates among-population genetic variation and reduces g^2 toward zero; (2) opposition between the levels of selection; and (3) a simple, additive one-gene basis for both individual and group fitness. Authors of these models did not appear to appreciate that each assumption, reasonable though it might be for mathematical analysis, favored individual selection over group selection. Wright (1959) (and to a lesser extent Lewontin [1974, p. 283]) had discussed how such simplifying assumptions were dual edged. They reduced dimensionality and increased tractability but at the same time limited the kinds of biological phenomena that could be modeled.

In the following sections, I will explain the reductionist assumptions of the models, our concerns about them, and the experiments we designed to address them. I began testing these models in my first year as an assistant professor and continued them in collaboration with Dave McCauley, who joined the lab in the fall of 1976 as a postdoctoral fellow. G. C. Williams had been a member of Dave's dissertation committee at Stony Brook, where most graduate students "knew better than to talk about group selection" (McCauley, pers. comm.) Together, Dave and I developed an experimental approach to investigating adaptive evolution in metapopulations using multilevel selection as our guide. At the same time, we figured out how to have a career in science that included undergraduate teaching and the training of doctoral students.

The Origin and Maintenance of Among-Group Variation

Phenotypic Variation among Populations

Selection requires phenotypic variation (Lewontin 1970b). Just as individual selection requires variation among the members of a population, group selection requires variation among populations. I will use the symbol $V(z)$ to represent the variance among individuals within a population, $V_{ave}(z)$ the metapopulation average of $V(z)$, and $V(Z)$, for the variance among population means, where the Z of a population is the average of the z's of its members. The total variance across a metapopulation is the sum of $V_{ave}(z)$ and $V(Z)$. Individual selection requires that $V_{ave}(z)$ be greater than 0; when it is zero, individuals are identical to one another in phenotype and there can be no selection. For the same reason, group selection requires that $V(Z)$ exceed zero. Theory allows us to draw a direct connection from extinction, colonization, and migration to the relative values of $V_{ave}(z)$ and $V(Z)$ (e.g., Slatkin 1985; Wade and McCauley 1988; Whitlock and McCauley 1990).

Under standard theory, a population is nothing more than the sum of the properties of its component individuals. Without interactions, the necessary relationship between $V(z)$ and $V(Z)$ poses a problem for group selection. When groups are established by sampling N founders at random from a large source population, $V(Z)$ should equal $V(z)/N$. Thus, $V(Z)$ is smaller than $V(z)$ by a factor of $(1/N)$. Also, smaller founding populations are better than larger ones for creating $V(Z)$ from $V(z)$ by sampling.

The creation of new populations by colonization involves sampling the colonists from a source population. The source population used in standard theory is a pool of migrants, some from every population in the metapopulation without regard for their mean phenotype, Z. Total variation, $V_{Total}(z)$, in the migrant pool equals the total phenotypic variance in the metapopulation, because the pool is created by mixing individuals from all populations. This is called *island-model migration*. With K colonists, the variation among the newly colonized populations, is only $V_{Total}(z)/K$. This means that colonization increases $V(Z)$ each generation by an amount equal to the fraction of new colonies in the metapopulation, c, times the variation among them, $V_{Total}(z)/K$. In natural populations, c seems to be small, ranging between 0.01 and 0.06 per generation (Wade and McCauley 1988). Yet, in my experiments, $V(Z)$ had increased much more rapidly and by much larger amounts than theory permitted.

Maintaining $V(Z)$ in the face of migration is even more difficult than creating it by sampling. The random migration of individuals between populations at a rate m ($m > 0$) destroys a portion of whatever variation sampling created. In standard theory migrants are sampled from the same pool as the colonists, which is the strongest mixing possible. This mixing reduces $V(Z)$ by $mV(Z)$. Thus, $V(Z)$ is increased by sampling during colonization ($+[c/K]$ $V_{Total}(z)$) but decreased by subsequent migration ($-mV(Z)$). The possibility for group selection ($V(Z) > 0$) depends on the fraction, (c/K), which governs the increase in $V(Z)$, and (m), which governs its decrease (Wade and McCauley 1988).

The forces increasing and decreasing $V(Z)$ can balance one another (Wright 1931). The ratio, $\{V(Z)/V_{Total}(z)\}$, which Wright called F_{ST}, reaches a constant equilibrium value, with island-model migration, of approximately $\{1/[4Nm+1]\}$ in a metapopulation without extinction and colonization. Even one migrant per population per generation (i.e., a value of m equal to $[1/N]$), reduces F_{ST} from 1.0 with no migration to only 0.20. When extinction, colonization, and a tendency for genetic relatives to emigrate together are added to Wright's model (Slatkin 1985; Wade and McCauley 1988; Whitlock and McCauley 1990), F_{ST} is substantially larger than Wright's equilibrium value. Nevertheless, in 1975, $\{1/[4Nm+1]\}$ was the standard used to evaluate the likelihood of group selection.

Whenever the number of colonists is less than the eventual size of a population (i.e., $K < N$), then most of g^2 is created by colonization (Wade and McCauley 1988; Whitlock and McCauley 1990). The equilibrium value of F_{ST} now depends upon the metapopulation reaching a "stable age" distribution of populations. In my experiments, K was 16 and N was as much as 30-fold larger. However, without migration, there was no force to decrease $V(Z)$. If we added migration to our experiments, $V(Z)$ would have to decrease.

If critics were correct, introducing very small amounts of migration would bring $V(Z)$ to zero and halt any response to group selection. We could test this prediction experimentally and the ancillary experiment (chapter 3) had partially allayed this concern. For critics, however, the important point was that a small but natural change in my experimental design—that is, adding migration, would radically alter the efficacy of group selection, diminishing the importance of our findings.

In summary, with sum-of-the-parts logic, $V(z)$, the phenotypic variation among individuals, should be several times larger than $V(Z)$, the variation among populations—a situation that greatly favors individual selection over group selection. My thesis observations challenged this logic: in the random treatment, I had observed $V(Z)$ to increase rapidly from 0 to 80%, well beyond its theoretical upper limit with a population size of 16 adults.

One explanation for the difference between my observation and the theoretical expectation was that small genetic differences created by sampling (colonization and drift) were amplified subsequently by ecological processes into much larger differences among populations. We also had to consider a more disappointing possibility—namely, that a population founded with 16 adult beetles was genetically much more like a population founded with only 1 or 2 beetles. If genetic differences arose only by random drift, then, according to theory, the $V(Z)$ would increase by about 3% per generation ($[1/2N_e] \sim [1/32] \sim 0.3$). The proportion could be higher if the parameter, N_e, was smaller than the actual number of founding adults, 16. In that case, my group selection might be closer to family selection than to Wright's interdemic selection. This possibility too required careful experimental study.

Fitness Variation among Populations

Selection also requires fitness variation. Once phenotypic differences exist, those differences must cause differences in births and deaths. For individual selection, the differences in z among individuals ($V(z) > 0$) must lead to differences in their survival and reproduction. For a metapopulation, the differences among populations ($V(Z) > 0$) must lead to some populations avoiding extinction longer than others or some sending out more colonists or migrants than others. Although there is no necessary relationship between the fitnesses of individuals within a population and the fitness of one population relative to another population (Lloyd 1988), standard theory assumed, yet again, a sum of the parts relationship: the fitness of a population equaled the average fitness of its members (Williams 1966). That is, if the fitness of the i-th member of a population is $w_i(z)$, then the fitness of the population, $W(z)$, equals the average, $\Sigma_i (w_i[z])/N$, where N is the number of its members. The variation among populations in fitness, $V(W)$, thus had to be small for the same reason that $V(Z)$ had to be small (above). Group selection among populations would necessarily be weaker than individual selection because $V(w)$ had to be larger than $V(W)$.

My experimental results did not conform to this expectation. The abundant and increasing variation in $V(W)$ indicated that population fitness was different from individual fitness. Measuring several of the primary characteristics of individuals had confirmed a lack of correspondence between $V(w)$ and $V(W)$ (Wade 1978a). But, where did the differences in offspring number among populations come from if not from differences in offspring numbers among their members? I did not yet have an answer.

Arguments against group selection in favor of individual selection also assumed that individual selection acted in the same way within every population. If, instead, selection acted in different directions in different populations, then individual selection could limit itself. Although individual selection acted within each population, when averaged across all populations, it might produce no net change at all. Under these circumstances, individual selection might locally be stronger than group selection, but the response to group selection could exceed the *average* response to individual selection.

Variation in the direction of natural selection from one population to another was viewed as a primary mechanism for maintaining genetic variation within natural populations (e.g., Hedrick et al. 1976). In my laboratory metapopulations, I had strictly controlled the major features of the environment (temperature, humidity, amount of resources), making sure they were the same for all populations. However, it was possible that the beetles created important ecological differences among populations by their own interactions and this variation could cause natural selection to vary in direction among demes within my metapopulations. To explore this possibility, we would have to devise methods for measuring individual selection within populations. This was a challenge. A flour beetle mates several times an hour with random partners; females randomly disperse eggs throughout the flour while tunneling; and, upon hatching, the feeding activities of the larvae completely mix the offspring of different beetles. To measure the fitness of one beetle out of 16, I would have to track each of these activities. To estimate the variance in individual fitness within a population, I would have to do it for every beetle! This tracking problem seemed impossible. But, without such information, how could I determine whether variation in the direction of natural selection from one deme to another was an important factor in my results?

Searching the literature, I found that Bateman (1948) had used a suite of visible mutations to uniquely assign offspring to each male and female parent in a population of fruit flies, *Drosophila melanogaster*, in order to measure the variance in individual fitness within a population. Unfortunately, we did not have a similar range of visible markers available in *Tribolium*. Fortunately, there were many other experiments to do. And, serendipitously, reading Bateman (1948) had further stimulated my interest in sexual selection.

Group Heritability

The last requirement of Darwinian logic is that the individual differences in phenotype, z, responsible for the individual differences in fitness, $w(z)$, must be transmissible to an individual's offspring (Lewontin 1970b). The problem for group selection was the inference from Fisher's Fundamental Theorem that natural selection eliminates heritable variation in fitness from a population. Hence, there should be very little *genetic* variation for fitness left in any population, natural or laboratory. Without heritable fitness variation within populations, neither colonization nor drift could create new heritable variation among them. Thus, the lower-level selective constraint on genetic variation for fitness negated the possibility of genetic—that is, heritable, variation at the higher level.

I knew from my random-extinction treatment that the observed population differences in fitness were heritable: (1) "offspring populations" founded with colonists from larger parent populations developed into larger populations, whereas offspring populations founded from lower productivity parents were of low productivity themselves; and (2) in both the high– and low–group selection treatments, there was a response to group selection. Why did we see g^2 at all, let alone such large values? I believed that g^2 must be composed of a kind of genetic variation different from that responsible for h^2 at the individual level. If there was genetic variation available to g^2 but not to h^2, it would explain not only the observed g^2 but also why variation in individual fitness persisted within our populations despite Fisher's Theorem.

"Additive" or sum-of-the-parts genetic variation is the only kind of genetic variation important for individual heritability, h^2. That is, parents might differ in fitness owing to differences among them in gene combinations, but only the average effects of their genes and not the special combinations themselves are transmissible to the offspring. R. A. Fisher lumped the gene-by-gene interaction variance with the nonheritable, environmental variance for this reason (see Wade 1985a; Goodnight and Wade 2000, p. 318). In contrast, *all* of the genetic variation, additive plus interactions plus social effects, contributes to g^2 and is available to group selection. An interaction effect is one that depends upon the particular configuration of genes within an individual. For example, return to the poker metaphor: the value of a three of spades depends upon the background of the other cards in a hand. It is more valuable in a hand surrounded by other spades (making a *flush*) but less valuable in a hand surrounded by a mixture of cards of other suits. This is what we mean by an *interaction effect*: the effect of a gene on z depends upon the *genetic background*, the other genes in the same individual. Thus, the same gene can have a large effect on body size in one individual but a small effect on body size in another, just as the three of spades can be valuable in one hand but worthless in another.

Just like the shuffling of cards prevents a player from "inheriting" the three of spades from the last hand *together with* the context that made it valuable, recombination separates individual genes from the genetic backgrounds that make them good or bad for fitness. This is why interactions do not contribute much to h^2. But, the interaction effects are heritable at the population level. By sampling or by local selection, the genetic background of one population becomes different from that of another. A group of colonists taken from a single population samples both a gene *and* its background, so the two are transmitted together as the colonists found a new group. This allows gene interactions to contribute to the resemblance of an offspring population to its parent population; it is why g^2 can be bigger than h^2.

It was also possible that my populations differed from one another in the genes responsible for social competition. Griffing (1967) was the first to investigate genetic, competitive social effects, which he called *associative effects*, but which are now called *indirect genetic effects* (Moore et al. 1998). In plants, especially crop plants, the best individual competitors tend to be those that take resources, like light, moisture, and nutrients, away from their neighbors. However, these most competitive individuals produce offspring of greater than average competitive ability. The offspring population may therefore have lower average fitness. This *negative* relationship between individual fitness and population fitness is the opposite of the positive sum-of-the-parts relationship assumed in the standard additive model. In a series of papers, Griffing showed how group selection was the only way to guarantee that the response to selection is positive when there are competitive interactions between individuals. If beetles created different social competitive environments, then those genes could contribute to g^2 but not to h^2 (cf. chapter 2).

For these reasons but with little quantitative understanding of them, we identified gene interactions and indirect genetic social effects as the most likely contributors to g^2 but not to h^2. Neither were prominent features of standard theory. Dave and I regarded indirect genetic effects as a better solution to the puzzle of rapid differentiation among populations with large N_e than gene interactions because of the biology of *Tribolium*. We could easily imagine that small initial differences in cannibalism and reproductive rate could be amplified by competition into very large differences in total offspring numbers. For gene interactions, we knew that Wright considered F_{ST} a one-gene measure that was "wholly inadequate" for describing the effects of drift on gene combinations. However, he had not extended F_{ST} to multiple genes, so we could not quantify the degree of the inadequacy he alluded to. In fact, for independent, neutral genes (those with no effect on fitness), the expected rate of drift for a two-gene combination was F_{ST}^2, which is a number *smaller* than Wright's F_{ST}, not larger. It was clear to us that "inadequate" did not necessarily mean "large."

We focused our theoretical efforts toward adding new kinds of genetic effects into the standard model to discover how to increase g^2 and make group selection more efficient. At that time, I did not appreciate that adding these interactions to existing models would necessarily reduce the efficacy of individual selection within populations. That understanding did not come until much later through the work of my student and collaborator Charles Goodnight.

Time Scales

In Nature, the extinction of a population is a much rarer event than the death of an individual. From the sum-of-the-parts perspective, extinction of a group was nothing more than the sum of the independent deaths of all a group's members. The membership of a population can turn over many, many times before it goes extinct, allowing individual selection repeated opportunities to cause adaptation. If episodes of group selection are rare but episodes of individual selection are abundant, then a single bout of group selection must be able to reverse the *cumulative* effect of several bouts of individual selection. This could be true only if g^2 exceeded h^2. Sum-of-the parts reasoning at every step in the Darwinian logic of standard theory left no possibility at all that one round of group selection could reverse the cumulative effect of multiple rounds of individual selection.

In our experiments, the lifetime of an individual and that of a population were commensurate by design, a circumstance not representative of most populations in Nature. Only in D. S. Wilson's trait-group models (D. S. Wilson 1980) was the lifetime of a group *less than* that of an individual, reversing the time scale problem. Like the absence of migration, the commensurate time scales of populations and individuals in our experiments limited the inferences we could draw from our laboratory model and apply to Nature.

Separating Selection from the Response to Selection

Just as individual selection might successfully oppose the lower level of gametic selection under some conditions, Dave and I believed there were conditions under which group selection might successfully oppose the lower level of individual selection. We discussed several ways that each of the concerns discussed above might be mitigated. If there were components of genetic variation available to group selection but not to individual selection, then it might be possible for the response to infrequent group selection to exceed the cumulative effect of more frequent individual selection. If there were sources of environmental variation reducing the heritability of individual differences but increasing differences among groups, then the total response to individual selection could be weakened at the same time that the response to group selection was strengthened. If the direction of individual selection at the gene level varied spatially among demes and temporally among generations, then its average effect might be small relative to a single bout of opposing group selection. It became clear from our discussions that some of our studies would have to be genetic while others would have to be ecological in order to investigate the possibilities we discussed.

My Chicago colleagues, R. Lande and S. Arnold, and I had found that evolution by natural selection could be better understood if it were considered as two distinct processes: selection and heritability. The strength of selection had been shown by breeders (Robertson 1966) to equal the covariance between phenotype and relative fitness, $Cov(z, w[z])$. When the $Cov(z, w[z]) > 0$, individuals with larger phenotypic values have higher relative fitnesses and those with smaller values have lower relative fitnesses. Moreover, the $Cov(z, w[z])$ equals $\Delta Z = Z' - Z$, the difference between the mean of the parent population after selection, Z′, and its mean before selection, Z. In order to compare one selection experiment with another, the difference, ΔZ, can be expressed in units of phenotypic standard deviations—that is, it is multiplied by $(1/[\ V(z)]^{1/2}]$. In this standardized form, one can determine where in the life cycle the selection is strongest; for which trait among a suite of traits it is strongest; and whether selection is stronger in one deme, ecosystem, or species than it is in another. Furthermore, the covariance between phenotype and fitness across an entire metapopulation, $Cov_{Total}(z, w[z])$, can be partitioned into two components (Price 1972): (1) average individual selection within populations, $Cov_{Individual}(z,w[z]) = \Sigma_j Cov_j(z,\ w[z])/T$; and (2) group selection among populations, $Cov_{Group}(Z, W[Z])$. This simple partitioning gave us a method for empirically comparing the relative strengths of individual and group selection.

How much of the change caused by selection acting on the parents, ΔZ, was transmitted to the next generation of offspring? This was the evolutionary change and it required knowledge of heritability because, for individual selection, the change from parent to offspring equaled $h^2 \Delta Z$. In order to compare *the genetic response* to individual and group selection, it was clear that we needed to measure heritability at both levels. Like selection, heritability can be expressed as a covariance—namely, the regression of offspring mean phenotype on parent mean. It can also be estimated as the variance among groups of relatives or as the realized or observed response to selection. In animal breeding, estimates of heritability prior to selection did not always coincide with the realized heritability, calculated from the actual response to artificial selection. (This was Griffing's motivation for recommending group selection over individual selection when breeding for yield.) If the animal breeders measured h^2 in three different ways, we thought that we should estimate g^2 in several different ways too. Group heritability was itself a new concept and we thought it particularly important to demonstrate that an estimate of g^2 prior to group selection was commensurate with the selection response observed after group selection.

Lastly, random drift in the absence of group selection caused g^2 to increase over time. If so, its rate of increase ought to be proportional to the strength of drift. Thus, we believed that $g^2(t)$ should be strongly affected by parameters, like m and N_e, as well as by extinction and colonization rates. Since many of the challenges to my doctoral results involved these same parameters, Dave and I planned a series of experimental studies to systematically vary N and m and observe their effect on $g^2(t)$.

For each experiment, we chose a wide range of parameter values. Some values we believed would allow $g^2(t)$ to increase over time but for other values we expected $g^2(t)$ would remain near zero. To connect $g^2(t)$ with the response to group selection—that is, to the "realized group heritability" of animal breeders, we planned to periodically impose group selection on the leftover beetles. This method also had the distinct advantage of addressing the time scale issue. We could find out whether or not a single episode of group selection, after several generations with no selection at all, could produce a large change in mean fitness. Our findings from this series of experiments are the topic of chapter 5.

5 Experimental Studies of Population Heritability

Teaching and Department Service

Once hired at Chicago as new faculty member, I was entirely on my own. Mentoring junior faculty was unheard of when I started in October of 1975. Research quality was the only criterion for hiring and tenure. Teaching responsibilities were not discussed during the job interview nor were they spelled out in offer letters, a practice that continued until 1991. Most of my colleagues complained openly about the burden of undergraduate teaching. I wanted to offer graduate and undergraduate courses in my specialty as my mentors, Park and Slatkin, had done. In fact, I developed my courses by copying them. I used Slatkin's syllabus to teach an undergraduate course, Introduction to Evolution. I modeled my graduate Population Biology and Field Ecology courses after Park's courses. My only teaching innovation was the addition of lectures on biometrics and experimental design to Field Ecology.

Because I was younger than the rest of the faculty and many graduate students, I felt an acute need to validate my new position by the quality of my teaching and mentoring. I had few mentoring opportunities because graduate students were not drawn to Chicago to study with an unpublished, unknown assistant professor. They came because of the famous tropical ecologists (D. Janzen and M. B. Lloyd), theoreticians (R. Levins and R. C. Lewontin), and behaviorists (Stuart and

Jeanne Altmann). When students did gravitate to my lab away from other labs, it was because of my teaching.

For Department service, I simply said yes to everything the chair asked me to do, a strategy that would be career suicide today. Fortunately, at that time, there were few standing faculty committees — no salary committee, space committee, or curriculum committee, and no mentoring, planning, facilities, animal care, website, or grievance committees. There were no jobs, so there were no search committees. Graduate student committees met only twice: for the proposal defense and for the thesis defense. Departments were not democracies and a chair's decision, made without much faculty input, was final. This was also a time when a majority of faculty smoked at faculty meetings. I realized quickly that faculty meetings were shorter when fewer people spoke; so, I kept my mouth shut, complained about teaching like everyone else, and threw myself into my research.

Effective Population Size in *Tribolium*

In theory, the strength of random genetic drift determined the rate of increase of g^2. The theory, however, was based on ideally behaving individuals, much like the ideally behaving "rational actors" of economic theory. Wright (1931, 1938) had emphasized, because real organisms behave in ways that deviate from the theoretical ideal, that a natural population's size, N, tended to be larger than its theoretical "effective size," N_e. According to theory, with an N of 16 beetles and no migration, g^2 should have increased by ~3% per generation. But, after 7 or 8 generations, g^2 was three or four times larger (0.65–0.83) than the value predicted by theory (0.21–0.25) assuming N_e was 16. If the effective size in my laboratory populations, N_e, was smaller, say 4 beetles instead of 16, then g^2 would be expect to increase by ~12.5% per generation and the difference between observation and theory would diminish considerably. If N_e were smaller still, the discrepancy would vanish. Moreover, if migration were added to our experimental protocols, any increment to g^2 would vanish. For these reasons, we had to estimate N_e and introduce migration among demes into our protocols to support the relevance of our findings.

What was the theoretically predicted effect of N_e and migration on g^2?

The Expected Genetic Differences between Populations

Wright created F-statistics to measure the among-deme fraction of the genetic variation. He called F_{ST} the *fixation index*, to emphasize its role in creating genetic differences among demes for the operation of interdemic selection, a type of group selection, in his Shifting Balance Theory (see chapter 9). He showed that fixations occur but are relatively rare when F_{ST} is less than 0.20 (Nm > 1), become common when F_{ST} is greater than 0.20, and predominate when it exceeds 0.33. For single genes, F_{ST} should have been directly proportional to our g^2. It could also be interpreted as the correlation between alleles drawn at random from within a population, similar to the r of kin selection. In an ideal population in the absence of selection, with N_e large and the rates of migration (m) and mutation (μ) small, Wright famously showed that the equilibrium value of F_{ST} is approximately $1/(4N_e[m + \mu] + 1)$

Many drew the following inference: "It is remarkable how little migration is required to prevent significant genetic divergence among subpopulations due to random genetic drift" (Hartl 1980, p. 195). They emphasized the genetic similarity among populations and dismissed Wright's argument that even small values of F_{ST} (01 to 0.05) represented evolutionarily important genetic differences. Critics of group selection went farther. "One migrant every other generation" (Nm = 0.5), or an F_{ST} equal to 0.33, was thought to be a necessary precondition for group selection (Wade 1978a).

Why did critics of group selection consider "one migrant every other generation" $(F_{ST} \geq 0.33)$ necessary for group selection? Why should group selection require genetic differentiation among demes in excess of those observed between species in the same genus (F_{ST} values near 0.10) or differences among genera (e.g., salamanders, Larson et al. 1984)?

The Origin of the Low Migration Condition for Group Selection

The belief that a very low migration rate was necessary began with Maynard Smith's (1964) "haystack model" of group selection (Wade 1978a). Maynard Smith imagined a population of mice distributed among haystacks. For genetic reasons, some mice were altruistic and self-sacrificing, willing to lay down their lives for their fellow mice; some other genetically different mice were selfish. Initially, altruistic and selfish mice were randomly distributed among haystacks, so that most haystacks contained mixtures of both types of mice. Maynard Smith imagined that, haystack by haystack, in a single generation, the selfish mice would take over and entirely eliminate the altruistic mice by individual selection. Even a single selfish mouse in a haystack filled with otherwise altruistic mice was sufficient. As a result, after individual selection within haystacks, only two kinds of haystacks remained: those with only altruist mice and those with only selfish mice. Only haystacks that, by chance, consisted entirely of altruistic mice at their founding could withstand the force of individual selection favoring selfish mice. These haystacks withstood the force of individual selection simply by not having the genetic variation that individual selection required. Thus, the existence of altruistic haystacks *after* individual selection demanded values of $F_{ST} \geq 0.33$, the region of F_{ST}-space where "fixation predominates" (Crow and Kimura 1970).

A value of F_{ST} near 0.33 makes a haystack of mice genetically intermediate between a half sib (0.25) and a full-sib family (0.50). At this value of F_{ST}, Maynard Smith's haystacks were not local populations at all but rather an array of families (chapter 2). Surprisingly, he used his theoretical results to draw a sharp distinction between group and kin (or family) selection!

Empirical Estimates of N_e and m

Dave and I considered the biological meaning of F_{ST} to be an unresolved *empirical* question. Did small values correspond to measurable and evolutionarily important differences among demes as Wright had argued or were they unimportant as claimed by critics of group selection?

Although there are now many indirect methods for inferring N_e from gene polymorphism, none were available for *Tribolium* in the mid-1970s. I tried to modify *Drosophila* electrophoresis protocols (agar gels) to study flour beetle allozymes as a way to measure polymorphism and infer N_e. Unfortunately, grinding up beetles released their toxic defensive chemicals (quinones) into solution, where they degraded allozyme activity, prohibiting staining. I also tried and failed at a new method, two-dimensional protein electrophoresis (O'Farrell 1975), which I thought might allow me to "see" the changes in gene combinations caused by group selection.

Dr. R. Selander at Cornell University, a renowned expert in electrophoresis, extended a generous invitation to visit his laboratory and, in a single day, he found levels of gene polymorphism in *Tribolium* comparable to or just a bit lower than those of *Drosophila*. However, the limited scope of this exploratory work did not allow us to place very precise confidence limits on the polymorphism estimates. Selander suggested that, with twenty more beetles per population and only one or two more variable allozymes, the overall picture of polymorphism in *Tribolium* might change substantially. I had neither the expertise nor the funds to extend this study on my own, so I turned to more direct methods of estimating N_e based on visible genetic markers, which Wright and his graduate students had pioneered in fruit flies (Kerr and Wright 1954).

The N_e of Beetle Populations

There had been only four previous experimental studies of the effects of random drift on gene frequencies, all conducted with fruit flies (Wright and Kerr 1954; Kerr and Wright 1954; Buri 1956; Dobzhansky and Pavlovsky 1957). Moreover, each had started with an initial gene frequency of 0.50 and used a fixed value of N. None had examined the relationship between N, m, and F_{ST}.

In my laboratory cultures, I had two single-gene, visible mutations: Chicago-black from Park's stock collection and c-SM black, a mutation that appeared in a population during my thesis. Each was semidominant and affected adult body color, allowing the three genotypes, b/b (black), $+/b$ (dark brown), and $+/+$ (reddish brown or "wild type"), to be sorted by eye. The experiments were simple (Crow and Morton 1955): a single b/b male or female was placed into a population of $+/+$ beetles (and vice versa). A month or so later, its $+/b$ offspring and those of other pairings $(+/+)$ were separated and counted. I could estimate the variation in offspring numbers for either sex in a population of any size or sex ratio. As a bonus, I could measure sexual selection and determine whether *Tribolium* males were more variable in offspring numbers than females as Bateman (1948) had found for fruit flies.

First, I estimated N_e for densities of 4, 8, 16, 24, and 48 adults, maintaining an equal sex ratio at every density (Wade 1980b). In the second study, I varied the sex ratio from 20% to 80% males in increments of 10% at two densities (N = 10 and 20 adults; Wade 1984c).

In both experiments, some b/b individuals produced significantly more or fewer offspring than expected. These deviations occurred at the lowest density and in the minority sex at the most extreme sex ratios. Importantly, there were equal numbers of + and − deviations, so there was no effect on average. This was different from the earlier experiments with flies (e.g., Wright and Kerr 1954). There, the markers themselves had very strong effects on viability, allowing natural selection to contaminate the estimates of drift.

In both studies (Wade 1980b, 1984c), males were more variable in fitness than females, as is typical of sexual selection based on male competition for mates (Bateman 1948; Wade 1979b; Shuster and Wade 2003). The ratio of (N_e/N) was 0.88 at low density and 0.92 at higher density (Wade 1984c). Furthermore, nearly all b/b beetles in both studies produced $+/b$ offspring, indicating that "migrant" beetles in our experiments represented real gene flow. Taken together, these results showed that $4N_em$ was much greater than 1 in the ancillary experiment of my thesis.

Estimating N_e was tedious and few were as interested in its value as I was. I measured it only because it was essential for placing the interpretation of our work on a more rigorous empirical foundation. Two and a half years of mind-numbingly boring work and the census of more than 250,000 beetles (Wade 1980b, 1984c) allowed us to add these eight words to our publications: "the ratio of N$_e$ to N is 0.88."

Dave noted another feature of *Tribolium* biology that could make N_e *larger* than N. In our metapopulation experiments, all females in a founding propagule were mated and carried the sperm of nonfounding males. That is, some offspring in our demes were sired by males left behind, an effect we called *sperm carry-over*. We estimated its magnitude by mating b/b females to b/b males and then placing those mated b/b females into groups of +/+ adults. Any b/b offspring of these females must have been sired by the b/b males left behind. We found that 3%–4% of offspring were sired by carried-over sperm (Wade and McCauley 1980). This effect boosted N_e to 92%–96% of N: *Tribolium were within a few percent of the theoretical ideal!*

These studies of N_e confirmed that our g^2 values were three to four times larger than theory predicted. We were now convinced that there was genetic variation not accounted for in theory that was contributing to g^2 but not to h^2. And, that this variation was beyond the reach of individual selection. Encouraged, we went ahead with studies of g^2 in metapopulations with larger values of N and with migration.

Genetic Differentiation versus Heritability

Using previous results with N = 16 and m = 0 as a baseline, Dave and I planned a series of three experiments: (1) metapopulations with larger values of N but no migration; (2) metapopulations with a constant N but with varying levels of migration; and finally (3) metapopulations with different combinations of both N and m.

We saw two complementary features of the relationship between N and g^2: differentiation and heritability. Considering drift alone, it was clear that an array of smaller populations should differentiate genetically faster than an array of larger populations. This implied that $g^2(N_{small})$ would be greater than $g^2(N_{large})$. On the other hand, larger colonizing propagules would be more "representative" genetic samples of their source populations. Thus, the heritability of population mean fitness should increase with the number of colonists, implying that $g^2(N_{large})$ would exceed $g^2(N_{small})$, the opposite relationship expected from drift alone. While theory addressed genetic divergence by drift with F statistics, it did not address the second feature, group heritability. Furthermore, F statistics were based entirely on single gene models. They were inadequate for describing differentiation owing to interacting gene combinations. Our experimental work headed into this "realm of inadequacy."

Experiment 1: Varying N without Migration

We explored a range of N values that bracketed the 16 adults of my thesis by creating metapopulations with 6, 12, 24, or 48 founding adults. With 48 adults, we believed that we would see no differences between populations at all, phenotypic or genetic. For the smaller N values, we worried that inbreeding depression might lower population mean fitness, so we planned to measure the magnitude of inbreeding depression midway through our experiment.

We also lowered the extinction rates, which had been very high in my thesis. Dave argued that high extinction rates would interfere with our ability to detect $V(Z)$ and g^2, since, in a few generations, all of our populations would be descended from just one or a few of the original founders. Influenced by Wright's comments on my doctoral work (chapter 9), I argued the opposite: extinction with recolonization was the "optimal condition" for generating g^2. We agreed on a compromise (Wade and McCauley 1980): no extinctions at all in one set of four metapopulations (the "Persistent" treatments) but 75% random extinctions in another four metapopulations (the "Extinction" treatments). Lastly, we eliminated the kind of "accidental" group selection that had occurred in the random-extinction treatment of my thesis work by taking exactly four offspring populations from each randomly selected parent.

For each of 14 generations, we censused the populations and computed $g^2(t)$. In the 4 Persistent metapopulations, $g^2(t)$ equaled the square of the parent-offspring population correlation. In the Extinction metapopulations, we estimated $g^2(t)$ using the variance components approach of my thesis. At generations four and seven, we estimated $g^2(t)$ in both ways by making several "extra" offspring populations from each population of the Persistent treatments.

We thought that, when g^2 was high, a single episode of group selection could have a large effect, something not considered in the group-selection debates. To explore this idea, we imposed high and low group selection on the "leftover" demes of all metapopulations at generation three, when values of $g^2(t)$ ranged from 30% to 70%. This "experiment within the experiment" gave us a third measure of $g^2(t)$, the "realized" group heritability (the ratio of the response to group selection divided by the group selection differential). In individual selection studies, it was routine to estimate heritability before selection, by parent-offspring regression or with half-sib designs, and compare the pre-selection estimate with the "realized" heritability, the actual heritability during a selection experiment. Realized heritability often differed among replicated selection lines as well as between up and down selection lines for the same trait. In fact, the largest deviations between estimated and realized heritability were found for fitness traits. The response to group selection gave us a way to measure realized $g^2(t)$ and demonstrate its relevance to the debate over the efficacy of group selection.

Results

In every one of the Persistent metapopulations, we found an increase over time in $V(Z)$, with the smallest increase associated with the largest N and the largest increase with the smallest N values (Wade and McCauley 1980, Fig. 2). Our results were in complete accord with theoretical predictions based on the effects of random drift on F_{ST}: the change in $V(Z)$ with time was inversely proportional to N_e.

It was the trends in $g^2(t)$ that surprised us. For all values of N, small and large, $g^2(t)$ was low in the early generations, increased over time, and then leveled off at values between 0.80 and 0.99 in the latter half of the experiment (generations 7 to 14). In the metapopulations with the largest N, we had expected an *immeasurably small* increase in $g^2(t)$ because F_{ST} should have increased only 1% per generation $(1/[2*48] \sim 0.01)$. How could the N-48 metapopulation and the N-6 metapopulation, which differed 8-fold in their F_{ST}'s, be equal in g^2? Clearly, F_{ST} and g^2 were measuring different things; group genetic differentiation was not the same as group heritability. Just as clearly, arguments against the efficacy of group selection based solely on F_{ST} were flawed, because g^2 could be large at the same time that F_{ST} was small.

In the Extinction metapopulations, trends in $V(Z)$ were much harder to detect and our estimates of $g^2(t)$ were much more variable from one generation to the next, just as Dave predicted they would be. Furthermore, again as Dave predicted, the $V(Z)$ tended to be greater for the Persistent than the

Extinction metapopulations. A high rate of random extinction did cause a net loss of heritable variation among populations in a finite metapopulation, just as random drift led to a net loss of heritable variation among individuals within a finite population. Nevertheless, average values of g^2 in the Extinction treatments reached values ranging from a low of 0.51 (N-24) to a high of 0.84 (N-48). Values of h^2 for individual traits in this range were considered high by animal breeders. Thus, despite an N_e *three times larger* than that used in my doctoral work (48 adults versus 16), we still observed very high values of g^2 for mean fitness.

We also found that a single episode of group selection produced a highly significant response in 6 of the 8 cases, and significant in at least one direction for every value of N_e. Like Wade (1977), the response in different directions was asymmetric: group selection for decreased numbers produced a greater response than group selection for increased numbers. More importantly, however, the magnitude of the response to a single bout of group selection was 30%–50% as large as the *cumulative response* to several consecutive episodes of group selection in my thesis. *Group selection did not have to act every generation to successfully oppose individual selection.* In the context of the group selection debates, this was truly a novel finding. This finding refuted the time-scale argument against the efficacy of group selection in the face of opposing individual selection.

Inbreeding and Population Fitness

We measured inbreeding depression in each of the Persistent metapopulations at generation 8. We created completely outbred populations with the same N and compared their productivity to that of the paired "inbred" experimental populations. In every case, the outbred populations were more productive than the corresponding inbred populations. The per-beetle inbreeding depression (the difference between the outbred and inbred means divided by the number of founders) was greatest when N was smallest (9.67 offspring per parent at N-6 adults versus 1.11 at N-48). Furthermore, $V(Z)$ was smaller among the outbred populations than it was for their inbred counterparts, as would be expected owing to the mixing necessary to create groups of outbred founders. These results supported our interpretation that (1) there were differences among populations in every metapopulation; (2) these differences were heritable; and (3) the genetic differences were owing in part to drift, which increased homozygosity (i.e., inbreeding), especially when the number of colonizing adults was small. The results not only conformed to theoretical expectations but they also ratified Slatkin's advice during my defense of my thesis proposal. His recommendation to use an N of 16 adults had been a terrific compromise; it allowed genetic differentiation while minimizing inbreeding depression. These inbreeding results were further evidence that we were studying an interaction between genetic variation and N_e, one that affected both group fitness and group heritability. Moreover, the results were entirely consistent with inbreeding and drift theory, validating the appropriateness of the *Tribolium* model system for studies of population genetic structure.

Although no one questioned the relevance of heritability to individual selection, our concept of "group heritability" was received with a great deal of skepticism when we submitted our findings to *Genetics*. One reviewer began his review in this way: "It is obvious from this paper why ecology is in such a semantic mess today. There have been so many terms in ecology that are essentially useless because they have been defined in an infinite variety of ways. Now Wade & McCauley (W & M) want to do the same thing to some perfectly good genetic terminology. I think that GENETICS should take a strong stand against such infringement by those who would alter the meaning of our concepts. I am referring, of course, to heritability. It is clear that W & M do not have a clear idea of what heritability really is (a common problem); in particular they have no idea what realized heritability is. If the science in the paper is acceptable, W & M should be required to completely rewrite it." The editor agreed, saying "Your measure g^2 is poorly defined and, I am afraid, actually misleading." Dave and I turned to *Evolution* rather than undermine our own empirical findings. Although one reviewer wrote "Bullshit" across one of our manuscript pages, it was accepted for publication after revision.

Because of reactions like these to our work, I did not submit grant proposals to the National Science Foundation early in my career. I worried that, if my proposal were distributed to 10–12 reviewers, per NSF practice, there would be a good chance that more than a few reviewers would become apoplectic reading about group selection. I was much more comfortable with the review process of the National Institutes of Health. Although there were formidable population geneticists on the NIH panel, like Drs. D. Hartl of Harvard and M. Feldman of Stanford, I believed that my ideas about epistasis, associative genetics, and population subdivision would receive a stringent and critical but fair and insightful review. Fortunately, that proved to be the case; indeed, we did better experiments informed by their comments on our proposals.

Experiment 2: Varying Migration While Keeping N Constant

In natural populations, migration most often occurs between neighboring populations, a pattern called *stepping-stone migration*. As the geographical distance between two populations increases, the genetic differences between them increase too, a phenomenon called *isolation by distance*. In contrast, most theoretical models, for tractability, assumed island-model migration, where populations exchange migrants without regard for physical distance; distant populations exchange just as many migrants with one another and just as often as neighboring populations. Island-model migration results in much greater genetic mixing of populations than stepping-stone migration. In fact, with island-model migration "a high degree of gene differentiation may occur only when there is practically no migration between populations" (Nei and Feldman 1972). For the same amount of migration, a stepping-stone model allows quite a bit more genetic variation between populations (i.e., higher F_{ST} values) than does an island model. Crow and Aoki (1982) showed that "one migrant . . . in the island model, is as effective as 3 or 4 when migration is restricted to neighboring groups in determining the equilibrium value of F."

To be conservative in our approach, we decided to experimentally investigate island-model migration even though this pattern of migration was clearly unfavorable to group selection (cf. Wade 1978b). If group selection worked with island-model migration, it would certainly work with stepping-stone migration. Still, it was hard for us to commit to a very large experimental effort when theory predicted that even small values of m would have devastating effects on F_{ST}.

In theory, island-model migration was equivalent in its effect on F_{ST} to migration into a group of populations from a common, external source maintained at a fixed gene frequency (Figure 5.1). We also investigated this equivalence by setting up treatments where the source of the migrants was other demes from within the metapopulation (MP, Figure 5.1A) and those where the migrants were taken from an external source, a stock population (BP, Figure 5.1B). For the MP source, we investigated four migration levels, 0%, 6.25%, 12.5%, and 25.0%, which spanned the range where group selection was considered possible (0%), unlikely (6.25%), or impossible (12.5% or 25%). For the BP source, we investigated only the two intermediate levels.

Each of the four MP and two BP metapopulations consisted of 35 populations (35 populations per treatment × 6 treatments = 210 populations total). Initially, each was founded with 16 adult beetles from our c-SM stock. In all

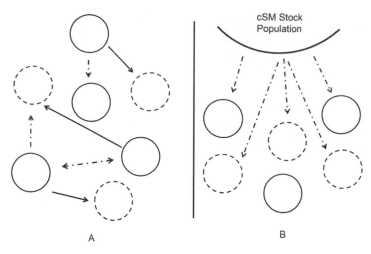

FIGURE 5.1 A schematic diagram of our protocol for the Migration Experiment: (A) the "metapopulation" model (MP); (B) the "boundary population" model (BP). Extinct populations or vacant habitats are indicated by dotted circles while extant populations are shown by solid circles. The large solid arrows represent colonization and the small dashed arrows indicate island-model migration. In (A), colonizing propagules (solid arrows) arise at random from across the metapopulation. In (B), colonizing propagules were drawn from the c-SM stock at the start of experiment as were all subsequent migrants (dotted arrows).

subsequent generations, seven populations were selected at random and, from each, we produced five "offspring" populations. The number of founding adults was constant at 16, but their population of origin varied depending on the treatment-specific level of migration. With our protocol, a group of colonists shared a "probability of common origin" of 0.71–0.82 (Whitlock and McCauley 1990).

In the MP-0% metapopulation, there were no migrants in any colonizing propagule, while the MP-6.25%, MP-12.5%, and MP-25.0% had 1, 2, or 4 migrants, respectively. The island-model migrants were chosen randomly from the 28 remaining populations in each metapopulation. For example, in the MP-12.5% treatment, two migrants were added to each propagule of 14 adults for a total of 70 migrants. After the migrants were added, every population in every treatment had 16 founding adults. The BP-6.2 5% and BP-12.5% treatments were handled like their MP counterparts, except that the migrants were from the c-SM stock population (Figure 5.1B).

As in Experiment 1 (above), we imposed single episodes of up or down group selection on the leftover beetles of each metapopulation at generations 1, 4, 7, and 9. This gave us six estimates of realized $g^2(t)$ (6 metapopulations × 2 directions of selection × 4 generations = 48 more episodes of group selection!). The five populations with the largest and the smallest numbers of adults were selected, and five "offspring" populations were made from each (10 parent populations × 6 metapopulations × 5 "offspring" per parent = 300 extra populations). Each offspring population was founded with a total of 16 adults, including migrants from the unselected populations, which reduced the strength of group selection by $(1-m)$.

We expected migration to have two effects: (1) it would reduce the among-population genetic variance available to group selection prior to selection; and (2) it would diminish the strength of group selection, because of the addition of unselected beetles to every selected population. The greater the migration rate, the larger would be the combined effect of *both* in preventing a response to group selection.

Results

Neither the number of migrants nor the source of migrants (MP vs BP) had a significant effect on W, population mean fitness. However, consistent with theory, the increase in *among-population* variance, $V(Z)$, over time was inversely proportional to the level of migration. In MP-0%, MP-6.25%, and BP-6.25%, $V(Z)$ increased significantly, while in those metapopulations with higher migration rates (MP-12.5%, MP-25%, BP-12.5%), it did not. This was qualitative agreement between theory and observation, for all levels of migration. However, the estimates of $g^2(t)$ were significantly greater than zero in every case for all levels of migration. Migration did not erase the capacity of populations either to differentiate from one another or to respond to group selection. Again, we had discovered that group genetic differentiation, estimated from F_{ST}, was different from group heritability.

Realized Group Heritability

The response to a single episode of group selection was significant in 34 of 48 cases — that is, fully 71% of the time. Trends in realized $g^2(t)$ as a function of migration rate were hard to discern. The fewest significant responses to group selection occurred in the first generation (5 of 12) and in those treatments with the higher rates of migration. For the lowest migration rates, 6.25% and 12.5%, there was significant heritable variation among populations and a significant response to group selection at every generation. Furthermore, across all migration rates, we saw a significant response to group selection several times in each treatment, including the MP-25%

TABLE 5.1 Average number of adult offspring produced per deme by 16 colonizers for the experimental high and low metapopulations relative to the unselected control metapopulations. In all cases, there was a significant response to group selection except with 25% migration (row 5).

Treatment	High	Control	Low	Significance
MP-0.00%	294	255	118	<0.001
MP-6.25%	226	173	151	<0.080
MP-12.50%	314	261	251	<0.050
MP-25.00%	239	215	240	N.S.
BP-6.25%	268	230	175	<0.001
BP-12.50%	268	240	207	<0.001

treatment with the highest level of migration. This response occurred despite the addition of migrants from the unselected populations and despite a high rate of random extinctions shown empirically (Wade and McCauley 1980) to be unfavorable conditions for detecting group selection.

The response to the episode of group selection imposed at generation 9 is illustrative. In Table 5.1 (after Wade 1982c, Table 3, p. 957), I report the means of the high–, random–, and low–group selection populations. *On average, despite migration, one episode of group selection created a difference in density of 61 beetles per deme between the High and Low metapopulations.*

In no case did island-migration homogenize populations to the extent predicted by theory, wherein "relatively low interdemic flow migration rates, on the order of 5% or less, would be required" for group selection to overcome opposing individual selection (Levin and Kilmer 1974; see also Maynard Smith 1964). Instead, our results for m and N together "indicated that population structure and intergroup selection may play a larger role in the evolution of populations than is generally acknowledged" (Wade 1982c). We decided to expand upon this conclusion by combining the separate studies of N and m into a single larger study varying both parameters at once.

Experiment 3: Varying both m and N

Informed by our previous findings, we increased N two-fold (from 48 adults to 96 adults) but decreased m by dropping the highest migration rate (25%).

Our design allowed us to determine which factor, N or m, was more important with respect to g^2. We set up 9 metapopulations, each with 15 populations (3 N values × 3 m values × 15 populations = 135 total populations). For N, we used 16, 48, or 96 adults to found populations. For island-model migration, we used values of 0%, 6.25% and 12.5%. Across this range of N and m, our *median* value of F_{ST}, even after 11 generations, would be only 0.06 in theory, far below the value (0.33) believed necessary for group selection (Maynard Smith 1964).

We varied the amount of flour (from 8 g to 24 g) and the size of our containers to accommodate the much larger number of founding beetles with respect to both volume and surface area. With 24 g of medium per vial instead of 8 g, the initial density of beetles ranged from 0.33 to 2.00 beetles per gram of flour, a very large ecological difference. To facilitate comparisons with our earlier results at 2 beetles per gram of medium, we founded four additional populations every other generation with 16 adults in 8 g of medium from a randomly chosen subset of populations from every N and m treatment. These uniform conditions of density permitted us to see the cumulative effects of population genetic structure rather than the demographic and density-dependent properties associated with a specific ratio of beetles to flour. This refinement to mitigate density differences was a lot of extra work, even if done only every other generation (9 metapopulations × 6 parent populations × 4 offspring populations = 216 extra populations). As an added bonus, however, these additional populations at a "uniform density" gave us a second way to estimate $g^2(t)$ while showing us how it might be affected by density-dependence.

We expected populations in the 96-12.5% treatment to be an upper limit to genetic differentiation. That is, they should experience little or no inbreeding owing to the combination of their large size and high migration rate. We thought we could safely attribute any decline in population mean fitness here to individual selection acting within populations.

Results

The differentiation among populations, $V(Z)$, increased significantly across generations, for all six metapopulations with the largest values of N (Wade and McCauley 1984). Within each level of migration rate, metapopulations with smaller N differentiated to a greater extent. However, for a given value of N, we did not see diminishing variation with increasing m as theory predicted. Part of our inability to detect this trend may have been statistical in that, where the theoretically predicted variation among metapopulations in F_{ST} was large (among N's within m), we detected the expected pattern but, where the predicted range of F_{ST} was much smaller (among m's within N), we did not.

In every metapopulation, $g^2(t)$ increased over time. Its largest increase occurred in the 16-0% metapopulation with the smallest N and no migration (m = 0). Its smallest increase occurred in the 96-12.5% metapopulation with the largest N and the highest migration rate (m = 0.125). Our findings refuted the theoretical claim that very small N and no migration were prerequisites for the existence of heritable variation among demes.

When we experimentally stripped away the density differences, using the uniform density populations, we found highly significant effects of both N and m on population mean fitness. There was a strong correlation of 0.75 (df = 38, P < 0.010) between F_{ST} and population mean fitness (McCauley and Wade 1981): the lowest values of F_{ST} had the highest value of W, a classic signature of inbreeding. (It is notable that we had to rear and census 1.25 × 10^6 beetles to get the 38 degrees of freedom for estimating this correlation.) Thus, we could account for half of the decline in population mean fitness by inbreeding (0.50 ~ [0.75]^2). The other half we attributed to uncontrolled individual selection within populations favoring behaviors like cannibalism. However, our ability to account for mean fitness in terms of genetic heterozygosity within populations told us little about the nature of the heritable variance among demes represented by $g^2(t)$.

Our data showed clearly that *Tribolium* populations were sensitive to very small values of F_{ST}, as Wright had argued. And, we could infer that, if genetic drift played a significant role in determining the fitness of populations with the island-model migration of theory, it would certainly do so with the stepping-stone migration characteristic of Nature. Overall, the results of our several studies together provided resounding and unequivocal empirical support for Wright's (1969, p. 293) assertion of that: "a rather small value of F [0.01–0.04] may be associated with a very considerable amount of differentiation among the subdivisions."

During this period, to relieve the relentless tedium of beetle lab work, Dave and I would often drive out to the old fields of the Cook County Forest Preserves where I taught my Field Ecology class. One July afternoon, staring at an expanse of Queen Anne's lace, Dave said, "You know, Mike, in ten days this field will be crawling with soldier beetles. Let's do something with them." On the spot, we planned a field project that became my first study of sexual selection: female mate choice in the soldier beetle, *Chaliognathus pennsylvanicus* (McCauley and Wade 1978). This study went from planning to press in only two months!

Development of New Theory Based on Our Experimental Findings

Our experimental findings with *Tribolium* prompted Slatkin (1981) to investigate how the mean of a trait with additive genetics would behave in a metapopulation with extinction and colonization. Results of his theory paralleled our empirical findings in several respects: (1) the total among-population variation in trait mean increased rapidly; (2) the increase was curvilinear over time; (3) $g^2(t)$ was only weakly dependent on N; and (4) migration among populations slowed but did not eliminate these features. His conclusions mirrored ours (Slatkin 1981, p. 869): "Taken together, these theoretical and experimental results should partially refute the common idea that group selection requires small deme sizes or complete isolation of demes to be effective."

Dave and I were impressed by the congruence between our laboratory observations and Slatkin's theoretical findings, but we were not convinced that our trait, population mean fitness, had an additive genetic basis like that assumed in the model. We thought that either gene-gene interactions, indirect genetic effects, or both influenced W. We also thought it possible that small differences in gene frequency among populations might be amplified or diminished by the highly interactive and density-dependent ecology of *Tribolium*. As a result, genetically similar populations might become very different in fitness. Conversely, genetically very different populations might have similar mean fitnesses (Wade 1978a). These were additional reasons to experimentally study the relationship between random genetic drift, individual traits, and population mean fitness.

Individual traits and Population Mean Fitness

Of all our findings, one stood out. In the Persistent treatment with the highest N (Experiment 2 above), a group of 48 colonists from one population produced nearly 500 adult offspring in a single generation, whereas the same number of colonists from another population in the same metapopulation produced merely 85 offspring. Both populations were derived from the c-SM stock without artificial selection of any kind and both had been propagated every generation at relatively large *N* without migration. This huge difference in the capacity for population growth had developed in only 14 generations. Dave and I singled out this "high population" (HP) and "low population" (LP) for further genetic and demographic study using methods from Dave's doctoral work (McCauley 1978).

We set up many new populations from the HP and LP source populations with varying numbers of initial founders (2, 8, 16, 24, or 48 adults). In one treatment, we reared all groups in 8 grams of medium, so there was a 24-fold change in density going from 2 founding beetles to 48. In a parallel series, we increased the volume of flour and the surface area as the numbers of founding beetles increased to maintain a constant density of 2 beetles per gram. No matter what the initial numbers or density, the differences between the HP and LP populations remained striking (McCauley and Wade 1980). Furthermore, when HP males were mated to LP females (or vice versa), offspring populations were intermediate in fitness, proving the differences had a genetic basis!

The *magnitude* of the difference between the HP and LP populations changed greatly with density and initial numbers. With a constant volume of flour, the average difference was the smallest at the lowest densities (mean difference of 80 adult offspring at <1 beetle per g of flour). The difference increased five-fold to 400 adult offspring at 2 beetles per g, and diminished somewhat, to "only" 250–280 offspring, at higher densities. With constant density, the difference in productivity between 48 HP and 48 LP adults was amazing: more than 1,000 adult offspring per population in a single generation! The density-dependent ecology of population growth clearly affected the magnitude of the fitness differences between the HP and LP populations, but in all cases a very large difference existed. This was a fascinating interaction between genetics and ecology, but we had no idea why it happened.

Interactions Caused Differences in Population Growth Rates

To determine why, we set up 40 new HP populations and 38 new LP populations, each with 48 adults in 8 g of standard medium. (The LP population did not produce enough adult beetles to set up 40 new populations.) At 6, 10, 16, 22, 28, 34, 40, and 46 days after founding, we sampled 5 HP and 3 LP populations at random and censused all life stages: eggs, small larvae, large larvae,

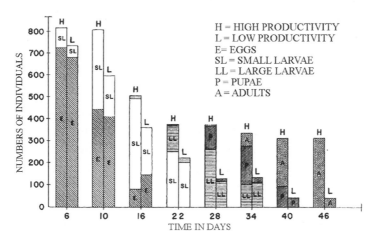

FIGURE 5.2 Diagram from McCauley and Wade (1980) shows the average numbers of each life stage in randomly sampled and sacrificed replicates of the high (HP) and low (LP) lines. 48 HP or LP adult beetles lay comparable numbers of eggs (eggs are shaded and labeled "E"; see days 6 to 16), but twice as many HP eggs hatch and survive to become small larvae (SL) owing to the greater incidence of cannibalism in the LP replicates. Despite much greater density, HP development rates were faster; note pupae appear a week earlier in HP than in LP. When larvae were reared alone, this difference vanished. Lastly, 85% of HP pupae survived and metamorphosed into adults but only 5% of the LP pupae. When reared alone, the difference in pupal mortality was only 2% instead of 80%. The only conclusion we could draw: cannibalism.

pupae, and adults. These census data were a statistically independent time series that allowed us to reconstruct the internal demography of the HP and LP populations, from the very beginning to the very end of a single generation. They allowed us to see *when* and *how* the large differences in mean fitness seen at 45 days developed.

We discovered that the seven-fold difference between the HP and LP population growth rates was the result of differences in the way the immature stages interacted with one another (see Figure 5.2). The HP and LP populations were similar in fecundity and viability. But, there was much less egg and pupal cannibalism in the HP population. In addition, HP larval development was insensitive to density, while LP development was delayed by density.

Drift had caused the HP and LP populations to become different in their social and ecological context. It was these indirect or associative genetic effects that were the source of the very large and heritable differences in W in our metapopulations. Where my tedious, earlier measurements on beetles from the high– and low–group selected lines (Wade 1977, 1978a) had failed to provide an explanation for population fitness, this experiment of Dave's had succeeded!

Inferences from the Laboratory to Nature

Taken together, our findings broadened the metapopulation structures where group selection might be important in Nature. First, our results had shown that the social and ecological context of competition was an important determinant of population fitness in *Tribolium*. Competition for resources, of course, is also the ubiquitous, driving engine of Darwinian natural selection. In most organisms, competition over territory, feeding sites, or nesting sites is commonplace. Secondly, the indirect effects of genes on competition were sensitive to group selection but not to individual selection. Because all individuals within a group experienced the same "average" level of competition, indirect genetic effects for competitive ability did not *differentially* affect individual fitness within demes. Indeed, in experiments in other labs, competition had not responded to direct artificial individual selection, a finding which led Futuyma (1970) to conclude, "The observations are consistent with the hypothesis that the genetic variance for traits affecting the outcome of competition is non-additive in large part, which suggests that some ecologically important characters may be subject to genetic homeostasis that prevents rapid response to environmental change." We had shown to the contrary that change in competitive context could be quite rapid as long the level of selection was the group and not the individual. Group selection was a more efficient way to change intraspecific competitive ability than was individual selection. The same genetics that prevented a rapid response to individual selection allowed a rapid response to group selection. Lastly, our interaction traits were the opposite of the "positive" social traits being discussed in kin selection theory. Instead of "altruism," which helped the group at the expense of the individual, ours were competition traits, such as cannibalism, that helped the individual at the expense of the group.

Many further analyses of our group-selection response were necessary before these inferences could stand on their own. Would group-selected gains disappear when group selection was relaxed? Did our group-selected populations show the theoretically expected trade-off between population growth rate and interspecific competitive ability predicted by ecological theory? Was the response to group selection dependent upon the hybrid nature of our c-SM stock or would we observe similar results using "natural" populations of *Tribolium*, recently collected from feed mills, farms, pantries, or pet stores? Would group selection work as well in two-species meta-communities as it had in our single-species metapopulations? Would group selection be effective if N varied from population to population instead of being held at constant value? Our experiments in pursuit of answers to these questions are found in the next chapter.

6　Population Ecology and Population Heritability

The Ecology of Group Selection

Our experiments in metapopulations had shown large increases in $g^2(t)$ across a substantially wider range of values of N and m than predicted by theory. Moreover, we had shown repeatedly that a single episode of group selection could produce a change in mean fitness as large or larger than those observed in my doctoral work. In experimental evolution, however, a successful selection response is only the first step of a research program. Breeders want to know whether or not high-performing stocks derived by artificial selection can be maintained without further selection. Often, the selective gains achieved for economically important traits come at the price of reduced fertility or viability. Moreover, maintaining selected gains often requires constant vigilance by the breeder because decay of selection response is common when individual selection is relaxed (Mather and Harrison 1949; Robertson and Reeve 1952; Lerner 1958; Falconer 1965). Selective gains are lost for two reasons: (1) opposing "natural" selection at the individual level within stocks takes over when artificial individual selection is relaxed; and (2) recombination breaks down artificially favored gene combinations built up by selection. Would opposing "natural" selection within our demes lead to a loss of group-selected gains when we stopped imposing group selection? And, did the population growth rates achieved by group selection lower other aspects of population fitness, such as competitive ability?

We had shown for some of our metapopulations that individual selection opposed group selection. We fully expected that our group-selected gains would be lost when individual selection continued unopposed. In particular, the loss of gains owing to recombination was of great concern. I believed that gene interactions (epistasis), long ignored by standard theory, were the source of the genetic variation contributing to g^2 but not to h^2. Once we halted group selection, the favorable gene combinations assembled by group selection would be broken apart by recombination; and, as a result, the response would disappear. The only way to test this possibility was to relax group selection and see what happened.

Successful artificial individual selection almost always produces unwanted by-products. The most common is a decline in individual fitness (Falconer 1965) owing to inbreeding and to "antagonistic pleiotropy," the spread of genes good for the target trait but bad for overall fitness (see Robertson's 1967 review of *Animal Breeding* in the *Annual Review of Genetics*). Inbreeding occurs over time as the members of the current selected population can be traced back to fewer and fewer ancestors in the starting population. The favored descendants are genetically similar to one another not only for the favored trait, but also for a host of other genes carried along in the selection process; they thus become inbred. Our group-selected populations in later generations similarly traced their ancestry back to fewer and fewer of the starting populations.

Antagonistic pleiotropy causes fitness to decline as genes with deleterious effects on fitness but positive effects on the selectively favored trait rise in frequency. This results in a reduction of total fitness as individual selection progresses. In fact, breeders account for such expected losses by revising downward predictions of the anticipated gain from selection. In our case, competitive ability was widely considered the "best" measure of total population fitness, better than any single-species population measure (Carson 1961) and better than our group-selected trait, population growth rate. Artificial individual selection for larger body size (Dawson 1967, 1968) had been shown to reduce competitive ability in *Tribolium*. Moreover, Park's experiments (1948, 1954) showed that population size achieved in single-species culture was a poor indicator of competitive ability in mixed-species culture. Thus, it was possible that group-selection on population growth rate might have decreased overall competitive fitness. For these reasons,

I believed it was imperative to assay the competitive ability of the group-selected populations.

Unlike domestic breeders, I also had the task of addressing the relationship between laboratory and field ecology. It was especially difficult to compare results from a *Tribolium* laboratory with Nature given the "mistrust by field workers for laboratory results" (Mertz and McCauley 1980, p. 95). MacArthur (1972) in his very influential book, *Geographical Ecology*, had dismissed the entire Park tradition of laboratory ecology with the disdainful label "bottle experiments." In a similar vein, Pianka (1974) in his graduate text, *Evolutionary Ecology*, had disparaged the findings of laboratory ecology as biased, because they were derived from the study of model organisms, all of which were "*r*-selected species." (That is, species selected for their rapid population growth in the lab.) Tropical ecologists, including several of my Chicago colleagues, also expressed a strong preference for "natural experiments" over laboratory experiments, despite their statistical limitations owing to a lack of random assignment and absence of controls. And, some natural experiments were viewed as better than others. For example, some colleagues were skeptical of my teaching Field Ecology in the local forest preserves. As one put it, "Why teach ecology in a sewer when you can teach it in the tropics?" In short, the conceptual climate challenged the legitimacy of drawing inferences about Nature from laboratory studies, let alone the use of laboratory studies to investigate group selection.

I had additional concerns. I had created the c-SM stock population to enhance the genetic variation for population fitness. And, I wondered whether any "natural" population of *Tribolium*, let alone of some other species, possessed comparable heritable variation for population growth rate. Along these same lines, single-species laboratory populations pale in complexity to natural populations embedded in multispecies communities. Natural populations experienced a much wider range of environments, biotic and abiotic, than the populations in our incubators. This environmental variation contributes to individual differences and makes h^2 smaller by reducing the proportion of variation that is genetic. In fact, heritabilities are almost always lower in the stressful environments of Nature than they are in the benign environments of the lab or greenhouse. Similarly, I expected that environmental variation at the population level, varying from locality to locality, would lower g^2. What if the environmental variation in Nature lowered h^2, but lowered g^2 more?

To investigate the effects of environmental variation on h^2 and g^2,

I would have to vary the environments of individuals and of populations, independently. Although I could easily put entire populations in different environments, in a laboratory microcosm, individual beetles tunneled, mixed, and mated together, so that each experienced, on average, the same environment as any other. I had no solution for this problem and moved on to address other questions. Fortunately, several years later, my student, Charles Goodnight (1985), devised a brilliant method for decoupling the individual from the populational component of environmental variation using the plant system *Arabidopsis thaliana* (see below). He could lower h^2 without affecting g^2 and vice versa. With *A. thaliana*, Charles could also impose individual and group selection independently and thereby investigate how different patterns of environmental variation affected the response to each level of selection. This was quite an exciting methodological advance, whose findings I will discuss toward the end of this chapter.

Previous studies had documented two sources of ecological variation in population growth rate in *Tribolium*: variation in founder number and competition with other species. In Nature, the number of founders varied from population to population across a metapopulation. In our experiments, we had held it constant: the same N for all populations in a single metapopulation. We had varied it between metapopulations—but not between demes within a single metapopulation. Both Park (1932) and I (Wade 1977) had shown that populations founded with a greater number of adults produced more offspring than those founded with smaller numbers. Variation in founder number was clearly an ecological source of growth rate variation that would lower g^2. Competition with other species also added variation to population growth rates (Neyman, Park, and Scott 1956); it too would lower g^2.

For these reasons, we introduced ecological factors, such as climate, competition, and founder number, into our experiments in order to move the lab closer to Nature. Moreover, it was important to find a field system that we could use to "calibrate" the laboratory against Nature. I discovered such a system while teaching Field Ecology. Willow trees throughout the Cook County forest preserves harbored the imported willow leaf beetle, *Plagiodera versicolora*, a species whose larvae lived in groups and were prone to cannibalism, just like flour beetles. I will discuss the decade of collaboration with my student and postdoc, Dr. Felix Breden, in field studies of this species in chapter 8.

Park and I discussed each of these issues at some length. Our discussions set priorities for my future work, particularly on the scope and size of simultaneously run experiments. Park would often end our discussions with the statement, "It is a lot of work, Michaeljohn. But it has to be done." He would then chuckle and add, "And, you will have to do it because I am retired!"

Many of the experiments planned with Park and described below appeared as single-authored publications. Most of my doctoral students worked on thesis projects in other systems and on other questions, independent of my beetle research. Their research enhanced the quality of my own, and diversified my experience with other questions and organisms, including plants. I would not have ventured into research in so many different areas of evolutionary biology without the atmosphere of creative inquiry fostered by Chicago postdocs and graduate students, mine as well as those of other faculty.

In the next several sections, I discuss the flour beetle experiments designed to address the effects of increased ecological realism on g^2 and on group selection.

Relaxing Group Selection

I decided to relax group selection for three years, a year longer than the time spent imposing group selection. Since the largest and smallest group-selection responses were expected to erode the fastest, I chose the 10 most productive populations from the high-group selection metapopulation and the 10 least productive from the low-group selection metapopulation. The top 10 demes averaged 256 adult offspring per 16 founding adults in 40 days, while the bottom 10 demes averaged a mere 18 adult offspring in the same time from the same number of founders—a 14-fold difference in population growth rate. How fast would this difference disappear without group selection?

I reared these 20 populations using Park's (1948, 1954) protocol for single-species control populations, censusing the adults and renewing the medium every 30 days. After almost three years without group selection, I restarted each of the 20 populations with 16 young adults to compare the growth rate of the restarted populations with that of their group-selected progenitors (Wade 1977, 1979a). Strictly speaking, I did not need to census these populations every month for three years to discover the effect of relaxing group selection—the restart data alone were sufficient for that purpose. However, the census data revealed striking differences in demographic patterns of the high and low populations, which were a correlated response to group selection on population growth rate.

Results

Relaxing group selection did *not* lead to the erosion of selected gains! The census of the top 10 populations exceeded that of the bottom 10 by more than three standard errors every generation for three years (see Figure 6.1). Even the least productive of the top 10 demes achieved a population size greater than the average of the bottom 10. Indeed, the top 10 were more productive than the most productive of the Park laboratory strains (see strain c-I in Park et al., 1964, Fig. 2). The evidence was clear and incontrovertible: the group-selected demes maintained their average characteristics for years after group selection was halted (Wade 1984a).

A direct comparison of offspring numbers between the original and restarted populations reinforced this conclusion. Productivity of the original top 10 averaged 236 adults per population at the first 60-day census, almost identical to their average of 242 adults per population achieved by their descendants nearly three years later. Similarly, the mean of the original bottom 10 was 86 at the first 60-day census, while the mean of their 10 restarts was 103, larger indeed but not significantly so. These data spoke clearly: *relaxation of group selection did not result in the erosion of selected gains either by recombination or by individual selection within the demes.*

The Correlated Response to Group Selection

The monthly census data revealed additional "correlated" responses of the population ecology to group selection (Wade 1984a). By a correlated response, I mean that another population-level trait changed, even though I did not select directly on or for that trait. The population ecology of beetles under the Park protocols follows a predictable pattern. A population tends to grow rapidly from a small initial number of adult founders, reaches a peak density, and then returns to an intermediate size as the beetle population dies back to a sustainable number determined by the amount of resources (Park 1954; Lloyd 1968; Mertz 1969).

The high and low groups had very different population ecologies. The peak number of adults for the top 10 was more than double that for the bottom 10 (271 versus 126). In addition, the top 10 achieved peak numbers much faster and were prone to outbreaks, large and sudden increases in adult numbers. Outbreaks had been studied in Park's lab before (Mertz 1969) and occurred "when populations [passed] through age configurations which favor reproduction but which are unfavorable for the cannibalistic self-limitation

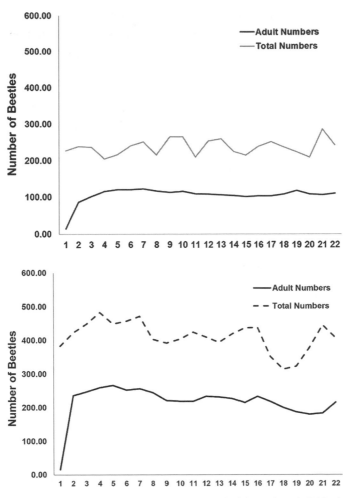

FIGURE 6.1 The Figures show the average 30-day census of adult numbers (solid line) and total numbers (dashed line; adults, pupae, and larvae) for the 10 low-group selected populations (top) and the 10 high-group selected populations (bottom; data from Wade 1984a).

of numbers" (Mertz 1969, p. 24). Outbreaks were three times more frequent in the fast-growing top 10 populations than they were in the slow-growing bottom 10. Where the top 10 were prone to outbreaks, the bottom 10 exhibited regular oscillations in the numbers of small and large larvae, pupae, and adults (see Wade 1984a, p. 408, Fig. 4). Such oscillations are characteristic of intense negative interactions between age classes, such as those resulting from the cannibalism of eggs by larvae or of pupae by adults (Lloyd 1968; Mertz 1969).

In a handful of generations, group selection had created, as a correlated response, two fundamentally different population ecologies representative of the textbook stereotypes of "r- and K-selected" species (MacArthur and Wilson 1967; Pianka 1970; here, r is the intrinsic population growth rate and K is the carrying capacity). The top 10 populations behaved like a classic "r-selected species." They combined a high population growth rate with a penchant for outbreaks. The bottom 10 mirrored a "K-selected species," with low population growth rates and density-dependent oscillations about their carrying capacity. This difference was particularly notable because several different *individual* selection experiments had attempted to create r and K populations. All had failed to generate even modest differences (Luckinbill 1978 with the bacterium *Escherichia coli*; Taylor and Condra 1980 with the fruit fly, *Drosophila pseudoobscura*). Taylor and Condra (1980, pp. 1190–1191) concluded, "It was expected that r-selected strains would have higher r, so this prediction was not met. . . . To the best of our ability to measure K, there was no difference in carrying capacity between the K-selected and r-selected lines." Another individual selection experiment by Mueller and Ayala (1981), also with *D. melanogaster*, was partly successful; it produced a difference of 1.0 offspring per pair at high density between r- and K-selected lines. The difference in offspring per pair produced by a group selection was more than 30 times larger! Group selection handily beat individual selection as a means of producing the iconic r and K populations central to ecological discussion then and now. Indirect genetic effects, which respond to group selection but not to individual selection, were responsible for our success (chapters 2 and 4).

Another striking feature of the relaxed selection lines was their continued, increasing differentiation from one another: the variation about the mean increased tremendously for both the high and low population groups. The variation in productivity, $V(W)$, of the top 10 populations increased by 4.6-fold ($V[W]$ equaled 1,782 originally but 8,120 among the restarted populations). The $V(W)$ among the bottom 10 increased 4.8-fold (1,112 versus 5,344). Were these differences among the high or low populations the result of three more years of random genetic drift or had individual selection acting in different directions within populations contributed to this differentiation?

My colleague Russ Lande (1976, 1977) had developed a method to answer this question, based on determining whether a group of populations was more different from one another than possible by genetic drift. I applied the Lande method to the high and low population groups, using conservative estimates of his model's parameters (see Wade 1984a, pp. 1042–1044). The observed increase in $V(W)$ was much too large to be accounted for by drift alone. This meant that individual selection was acting within some populations to make them larger but within others to make them smaller. *Thus, it was not the absence of individual selection but rather its lack of a coherent direction that preserved the group-selected gains during the years without group selection. This diversifying within-group selection was strong enough to increase the variation among populations nearly five-fold!* This result reinforced my speculation that heterogeneity in the direction of individual selection could contribute to g^2 and thereby to group selection.

Taken together, these results laid to rest one of my deepest concerns. Differences in growth rate created by group selection persisted in its absence. Individual selection within populations had not erased the group-selected gains. Furthermore, the high and low populations were different in many aspects of their population ecology in addition to the growth rate differences caused by group selection. The high populations reached a higher carrying capacity, did so faster, and were prone to outbreaks, while the low populations achieved a much lower K and exhibited regular oscillations about it, driven by competitive interactions.

Moreover, population differentiation had not reached its maximum under disruptive group selection. The high and low population groups differentiated several-fold more in the years without group selection, just as Dave and I had observed in our metapopulation studies of N and m.

Competitive Ability and Population Growth Rate

An ecological trade-off is a fundamental assumption within r- and K-selection theory. It assumes that a species cannot simultaneously maximize its interspecific competitive ability and its population growth rate (e.g., MacArthur and Wilson 1967; Pianka 1978). My group-selected populations with very large differences in r and K were tailor-made for testing the existence of this trade-off. According to theory, the high-r populations would be poor competitors, but the low-r populations should be good competitors. Park enjoyed using my group-selected populations in a *Tribolium* competition study because it linked his ecological work to my genetic work.

I assayed the competitive ability of the high–, low–, control, and random group selected populations from the ancillary experiment of my thesis (Wade 1977). From each of those 216 populations (4 selection treatments × 3 levels of migration × 18 populations) and the c-SM stock population, I established one single-species (SS) population with 8 males and 8 females and

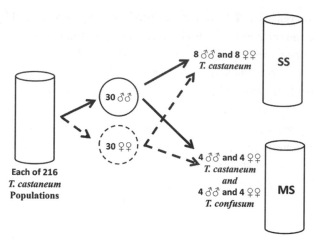

FIGURE 6.2 This is a schematic representation of the experiment designed to investigate the existence of the trade-off between population growth rate and competitive ability assumed by *r* and *K* theory. See text for further details.

one mixed-species (MS) population with 4 males and 4 females together with a similar number of beetles from a competing species, *T. confusum* (see Figure 6.2). By creating an SS and an MS population in this way, I was able to use a more powerful method of statistical analysis based on paired comparisons (Wade 1980c).

Après Park, I also set up 18 SS populations of the competitor species, *T. confusum*, for a total of 486 populations. I never renewed the flour in these populations, guaranteeing that each would become extinct within a year. If I had renewed the medium every 30 days as Park did, it would have taken 3–7 years for every MS population to resolve itself into a winning and a losing species. I examined each SS and MS population daily for extinction of either species. When I observed an extinction event, I censused the numbers of live and dead adults of each of the two species.

In the SS populations, the patterns observed in my doctoral work were manifest again. The high–group selected populations produced 45% more total adults than the controls (447 vs 308 adults) and the low–group selected populations 25% fewer adults than the controls (231 vs 308 adults). Not surprisingly, time to extinction was inversely correlated with productivity: faster growing populations went extinct sooner than more slowly growing populations. The negative correlation between productivity and time to extinction was very strong, ranging from –0.424 to –0.890. These results mirrored and replicated the findings of Nathanson (1975), whose research had given me the original idea for a group selection experiment. In fact, some of the correlations I observed were larger (i.e., more negative) than those seen in Nathanson (1975).

The data confirmed the foundational assumption of r and K theory: high SS population growth was correlated with low MS competitive ability. A population with high fitness in SS culture had low fitness in MS culture. The slowest growing SS populations were the best competitors because they had the greatest negative effect on the growth rate of the competing species in the corresponding MS populations. The high–group selected populations had the highest SS fitness, while the low–group selected populations had the highest MS fitness. Notably, the high competitive fitness of these slow-growing MS populations ruled out inbreeding depression as the cause of their low r in SS populations.

Competition prolonged the time to extinction of the most productive populations but shortened extinction time for the least productive populations (Wade 1980c), showing that the presence or absence of a competing species changed the mean population fitness. Consequently, a change in ecological context, such as the presence or absence of a competitor, could change the direction of group selection, favoring high growth rate in one circumstance but lower growth rate in another. This study supported the inference that the ecology of competitive exclusion as well as local resource limitation may play important roles in group selection in Nature.

Variation for Fitness in a "Wild" Population of *T. castaneum*

I needed to collect "wild" beetles to compare the natural variation in fitness with that of the c-SM stock, engineered to have genetic variability in fitness. Of the many possible sources of "wild" beetles, pantries, supermarkets, feed mills, and graneries, I chose a pantry population. I thought that a pantry population might represent a typical colonization event. And, because of its smaller size relative to the larger wild populations found in feed mills or supermarkets, it would be conservative with respect to g^2, harboring less genetic variation. Serendipitously, my technician, Ms. Ora Lee Lucas, vacationed in Little Rock, Arkansas, and returned with a bag of heavily infested rice from a relative's pantry. We collected over 100 adult beetles from it and used them to found the c-ARK population—our "wild" *T. castaneum*.

We established 20 populations, each with 25 c-ARK adults. We censused and transferred them to fresh medium in the same manner as the "control" metapopulation of Wade (1977) and the "Persistent" metapopulations of Wade and McCauley (1980). Over the next 2 years, group heritability, g^2, among the c-ARK populations reached +0.73 — a level of genetic differentiation comparable to those studies using the c-SM population.

At that point, when g^2 was high, I introduced each c-ARK population to less favorable and more stressful environmental conditions to investigate their effects on group heritability. The stressful environments were combinations of poor climate (27°C and 22% relative humidity) and interspecific competition (the presence of a competitor, *T. confusum*). I imposed 6 different environments: (1) a temperate-wet environment (29°C, 70% relative humidity) with no competitors; (2) a cold-dry environment (27°C, 22% relative humidity) with no competitors; (3) the same climate as (1) but with competition from 20 adult beetles of the b-SM strain of *T. confusum*; (4) the same climate as (1) but with competition from the b-I "Park" strain of *T. confusum*; (5) the same climate as (2) with 20 b-SM beetles; and (6) the same climate as (2) with 20 b-I beetles. The two temperature-humidity climates were the extremes of the range investigated in Park (1954) and the competitors were well-studied laboratory strains of *T. confusum*, known to differ in their population growth rates and competitive abilities when pitted against *T. castaneum*. Population b-I had a higher single-species growth rate but was a poorer competitor than b-SM.

Results

The census data provided evidence confirming that environments 2–6 were more stressful than the standard lab conditions (Wade 1990). In the cold-dry environment without competitors, 20 c-Ark adults produced only 77% as many offspring as they did when reared under standard conditions (mean of 57.3 versus 74.6 adult offspring). The addition of competitors was more stressful. The presence of a competitor reduced *T. castaneum* offspring numbers by 64%. The value of g^2 was smallest in the harshest environment but always greater than +0.34. As with h^2, the heritability of individual traits, stressful environments reduced but did not eliminate g^2.

The interaction between climate and competition was highly significant ($P < 0.035$). And, surprisingly, different populations responded to the 6 environments in different ways as evidenced by a highly significant three-way interaction of population × climate × competition ($P < 0.005$). Some populations were better suited to a colder climate than others, while others were better competitors. Furthermore, competitive ability depended not only on the c-ARK population of origin but also on the genetic identity of the competitor, *T. confusum*! Some c-ARK populations were better at competition with b-I than they were with b-SM and vice versa. Likewise, some were better competitors in one climate but poorer competitors in another. *This amazing amount of heritable variation in ecology was present within a small pantry population and revealed by random genetic drift in our laboratory metapopulation!*

Variation in Founder Number: Separating the Ecology and Genetics of g^2

In theoretical treatments of fluctuating N, the focus was on its *genetic* effects, while the simultaneous *ecological* effects had not been investigated. Variation in N causes nonheritable, ecological variation among demes in productivity, since a large number of founders produced more offspring than a small number of founders (Wade 1977, 1980b, 1984c; McCauley and Wade 1980). Therefore, fluctuation in founder numbers, variation in N among demes, should lower g^2 by adding nongenetic, ecological variation to population growth rate.

The well-known genetic effect of variation in N was an increase in the intensity of random drift. "Of all factors that affect the effective size, none is more important than fluctuations in the actual population size," (Gillespie 1998, p. 35; Wright 1969). According to evolutionary genetic theory, varying N should increase g^2 because that variation would strengthen random drift.

In short, the genetic and ecological effects of variations in N were predicted to have opposite effects on g^2. And, despite its acknowledged importance, fluctuating population size and random drift had not yet been experimentally studied. Even today, the absence of such data has been cited as a critical gap in our understanding of conservation genetics (Hedrick and Kalinowski 2000).

There was a further complication: our previous work had demonstrated that the ecological and genetic effects *interacted* with one another (McCauley and Wade 1980; Wade 1984c, 1990). In McCauley and Wade (1980), we had shown that the growth rate difference between the HP and LP demes changed with founder density because each population responded differently to changing founder density. In Wade (1990), I had shown that c-ARK populations responded in different ways to changes in temperature and competition. Thus, I expected that the *relative magnitudes* of the opposing ecological and genetic effects on g^2 would change over time because the ecological effect itself had a genetic component. Such interactions had not been considered at that time and have only recently begun to be explored in theory (Agrawal et al. 2001; Wolf et al. 2002). Because it is unlikely that such interactions between genetics and ecology are unique to flour beetles, this is another reason that *experimental* studies of metapopulations with spatial and temporal fluctuations in N should be conducted.

I designed the variable-N experiment so that I could separate total variation in population growth rate, $V(W)$, into its genetic and ecological components, in the same way that variation in individual traits is separated into components of nature (genetic) and nurture (environment). In the variable N treatments, I could partition the genetic and ecological influences on variability into separate effects by using founder number as a covariate in the analysis of population growth rate. Furthermore, I predicted that the magnitude of the ecological effect on g^2 would increase as the magnitude of the fluctuations in N increased.

I set up 6 experimental metapopulations, varying both N (10 or 30) and the fluctuations about N—that is, V_N (0, 10, or 30). I used a program written by Dr. C. Goodnight so that the variation in N would conform to a normal, bell-shaped distribution. Each generation, I plugged the desired N and V_N into the program and it produced a random value of N for each deme in the metapopulation. At generations 2 and 9, I imposed group selection on the leftover populations as Dave and I had done before (Wade and McCauley 1980).

At the end of the experiment, I measured the *interaction* between ecological effects and genetic effects on g^2 using the design from McCauley and Wade (1980). From each of the three small N metapopulations (10-0, 10-10, and 10-30), I randomly selected five demes. From each one, I set up three replicate populations with 2, 4, 8, 12, or 16 young adults in 8 grams of medium and censused them after 50 days (15 populations × 3 replicates × 5 values of N = 225 total populations). This range of founding adults spanned the founder numbers achieved by the fluctuations about N imposed by protocol during the main experiment. For the larger N metapopulations, I selected 5 populations at random from only one of the three metapopulations (30-0). From each of those, I set up three replicate offspring populations with 15, 20, 25, 30, and 35 founding adults in 8 grams of medium.

Results

The variations in N had a much larger effect when N was small than when N was large (Table 6.1). The harmonic mean for the N-10 metapopulations was an 18.4% reduction in N when the variance was 10, and a 39.5% reduction

TABLE 6.1 The harmonic mean values of population size realized in the variable N experiment.

Teatment (N, V)	Generations					
	1	3	6	9	12	15
(10, 0)	10.00	10.00	10.00	10.00	10.00	10.00
(10, 10)	9.20	8.53	8.48	8.33	7.99	8.16
(10, 30)	8.82	8.47	6.64	6.24	5.78	6.05
(30, 0)	30.00	30.00	30.00	30.00	30.00	30.00
(3r0, 10)	29.85	29.59	29.40	29.44	29.38	29.28
(30, 30)	29.60	29.08	29.04	28.98	29.78	28.57

when the variance was 30. The corresponding reductions in the N-30 meta-populations were only 2.4% and 4.8%, respectively.

These values of N correspond to the theoretically expected F_{ST} values depicted in the Figure 6.3 (below). Clearly, the expected genetic effects of fluctuations in N were large when N was 10, but negligible when N was 30. Indeed, the three lines for (30, 0), (30, 10), and (30, 30) lie right on top of one another in Figure 6.3 and are essentially indistinguishable.

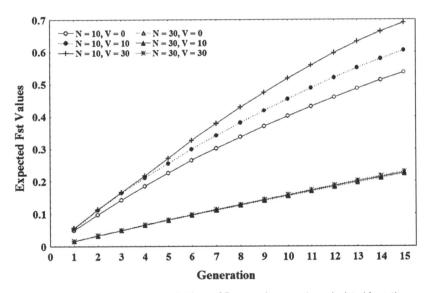

FIGURE 6.3 The theoretically expected values of F_{ST} at each generation calculated from the harmonic mean distribution of variations about N for the three metapopulations with an N of 10 ([10, 0], [10, 10], and [10,30]) and the three metapopulations with an N of 30 ([30, 0], [30, 10], and [30,30]). Note that variations in N about a mean of 10 increase the expected F_{ST}, but variations of the same magnitude about a mean N of 30 have a negligible effect on F_{ST}.

The variations in population growth rate were proportional to the theoretically expected values of F_{ST} (Figure 6.4A $N = 30$; Figure 6.4B $N = 10$), a finding similar to that of McCauley and Wade (1981). Here, an increase in F_{ST} of 0.10 corresponded to an increase of 0.05 standard deviations around the mean population growth rate, expressed as the natural log of total fitness, W, or $Ln(W)$.

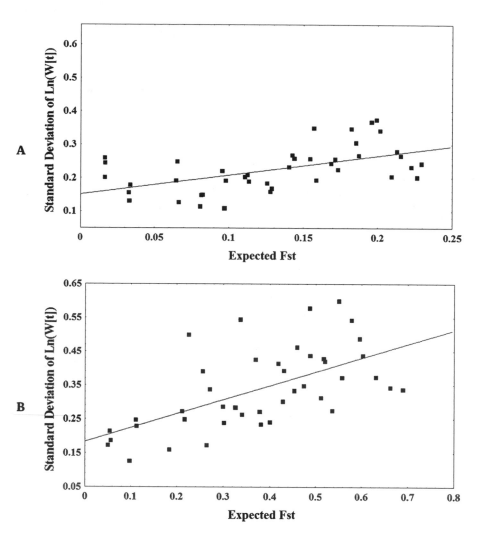

FIGURE 6.4 *A*, The variation in population fitness (S.D. of $Ln[W_t]$) regressed on the expected value of F_{ST} at each generation for the (30, 0), (30, 10), and (30, 30) metapopulations. *B*, The variation in population fitness (S.D. of $Ln[W_t]$) regressed on the expected value of F_{ST} at each generation for the (10, 0), (10, 10), and (10, 30) metapopulations.

The separate ecological and genetic effects on $g^2(t)$ revealed by covariance analysis were complex (Figure 6.5 and Figure 6.6). First, $g^2(t)$ increased with time for all four metapopulations with variable founding numbers and the largest increases were occurring at the smallest average N (10). This was not surprising, as we had seen this several times before: small demes di-

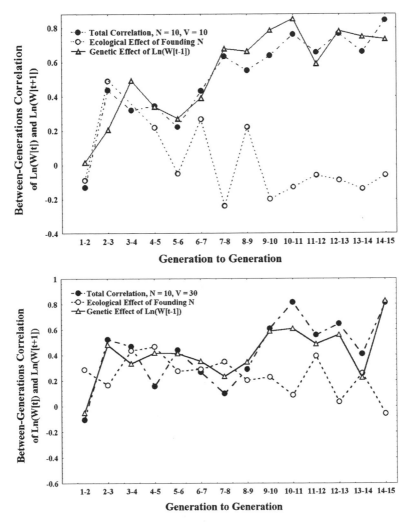

FIGURE 6.5 The separate ecological and genetic effects of founder numbers on population fitness (Ln[W_t]) for the (N 10, V 10) metapopulation (top) and the (N 10, V 30) metapopulation (bottom).

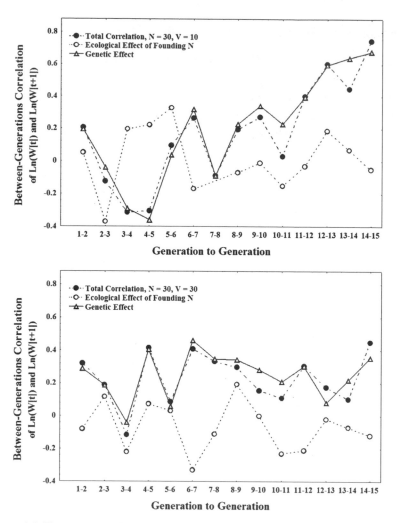

FIGURE 6.6 The separate ecological and genetic effects of founder numbers on population fitness (Ln[W_t]) for the (N 30, V 10) metapopulation (top) and the (N 30, V 30) metapopulation (bottom).

verged from one another faster than large demes. Second, the fraction of $g^2(t)$ owing to the genetic component increased over time in all cases, with larger increases in the metapopulations with smaller average N. Third, the genetic component tended to be larger than the ecological one in all cases, especially in the later generations. Lastly, in all cases, the ecological effect was greatest in the early generations, oscillated during the intermediate generations, and then settled down to small values near zero in the later generations. Differently put, as time went by, the ecological effects of variable numbers of founders was subsumed by the increasing genetic differences among the demes.

Results of Group Selection with Variable N

The response to group selection on the leftover populations was positive in nine of the twelve cases at generation three, significantly so for two and borderline significant for two others. These significant cases were the ones with the largest predicted values of F_{ST} and the largest observed values of g^2. At generation nine, the response to group selection was positive in eight of the twelve cases but significant only for three of them, 10-0, 10-10, and 30-0. These responses did not correlate well with F_{ST} or g^2, probably because the values of F_{ST} and g^2 were uniformly high for all metapopulations by this time. Combining the analysis across the two group selection experiments, there was a highly significant response to group selection with the smaller N (N-10: Chi-square = 20.026, df = 6, P < 0.003) and a nearly significant response with the larger N (N-30: Chi-square = 12.210, df = 6, P < 0.058).

The Interaction of Ecology and Genetics

There was highly significant genetic variation in the relationship between founding N and productivity in all four metapopulations tested. In two of the four metapopulations, I found a significant genotype-by-environment interaction (G × E) for population mean fitness, where genotype was the "parent population" and the environment was N, the founder density. There not only was genetic variation from one population to another in the relationship between founding N and productivity, but also small values of N from some populations were more productive than larger values of N from others and vice versa. *These changes in the relationship between N and population growth rate were another correlated ecological genetic response to random genetic drift acting on demes in a metapopulation, similar to the tendency toward outbreaks in the high–group selected populations or the density-dependent oscillations in the low–group selected populations.*

End of the McCauley Era

The Slatkin-Wade NIH grant, which had supported two and a half years of very fruitful collaboration with Dave McCauley, was expiring at summer's end 1979. I had written and submitted a renewal, but continued funding was as uncertain then as it is now. Furthermore, I would not learn the fate of my proposal until mid-July, putting Dave in a precarious position, since most fall academic positions would be filled before the start of summer. That spring, the University of Virginia (UVA) advertised Mountain Lake Summer Postdoctoral Fellowships, in an effort to boost the research profile of its field station. Dave knew of Mountain Lake and its summer entomology course from his time at Stony Brook. He applied for their summer postdoc in case the grant was not funded and in order to establish greater independence from me through field research. Dave enjoyed the field more than the lab. He had initiated our paper on mate choice in soldier beetles (McCauley and Wade 1978) and collected a large sample of mating and non-mating milkweed beetles, *Tetraopes tetraophthalmus*, (McCauley 1979) in New York. He based his Mountain Lake research proposal on those data, proposing to study stabilizing selection and sexual selection on body size. He was awarded the Mountain Lake fellowship and, later that summer, the station director offered him a temporary Visiting Assistant Professorship at UVA, with the possibility of it becoming a tenure-track position. Dave accepted this offer instead of returning to my lab, even though the grant had been awarded. The McCauley Era in the Wade lab was over (although our collaboration has continued for more than 35 years).

The pain of Dave's leaving was eased by the arrival of a very creative and somewhat manic cohort of Chicago graduate students, including Charles Goodnight, I. Lorraine Heisler, Felix Breden, Steve Tonsor, Sue Kalisz, Sam Scheiner, and H. Bradley Shaffer. A series of postdocs, Drs. Lori Stevens (from UIC), Felix Breden (from Chicago), and Chris Boake (from Cornell), replaced Dave as research associates on my renewed NIH support. In his doctoral research, Charles tackled a number of questions about variable environments and the levels of selection that Dave and I had found to be beyond the scope of the *Tribolium* system.

Goodnight's Thesis and the Scale of Environmental Variation

Charles used leaf area in the mustard, *Arabidopsis thaliana*, as the focal trait for his doctoral research on group and individual selection in a variable environment. His environmental source of variation was shading, which affects plant growth. With different layerings of nylon netting, Charles held the mean level of shading constant across a metapopulation while varying the level of shading among individuals within demes or among demes within a metapopulation. He imposed three types of environmental variation: varia-

		Between Deme Environment Variable		Between Deme Environment Constant	
		W/in Deme Environment Variable	W/in Deme Environment Constant	W/in Deme Environment Variable	W/in Deme Environment Constant
Group Selection For High Leaf Area	Ind. Sel. For High Leaf Area				
	Ind. Sel. For Low Leaf Area				
	No Ind. Sel.				
Group Selection For Low Leaf Area	Ind. Sel. For High Leaf Area				
	Ind. Sel. For Low Leaf Area				
	No Ind. Sel.				
No Group Selection	Ind. Sel. For High Leaf Area				
	Ind. Sel. For Low Leaf Area				
	No Ind. Sel.				

FIGURE 6.7 This schematic representation depicts Goodnight's doctoral thesis experiment investigating the interaction of different scales of environmental variation on individual and group selection.

tion only within demes, variation only among demes, and a combination of both. He reasoned that variation within demes should reduce the h^2 of leaf area but have little effect on g^2. Conversely, variation in shading only among demes should reduce g^2 but not h^2. A combination of shading within and among demes should reduce both. His was an heroic effort, in terms of both conception and execution (Figure 6.7). It involved 3 types of selection (up, down, and random), at 2 different levels (individual and group), with 2 types of shading (within and among demes) of 2 types (constant or variable), for a total of 36 different shaded metapopulations ([3 group selection × 3 individual selection] × [2 types of shading × 2 levels of shading] = 36 treatments with 9 demes per metapopulation). It remains one of the most beautifully designed metapopulation experiments in multilevel selection studies.

Charles' analysis revealed a rapid and significant response to group selection for increased and decreased leaf area. In contrast, individual selection for increased leaf area resulted in the opposite response: a decrease in leaf area relative to the unselected controls. Since plants with larger leaves achieve their size at the expense of their neighbors, individual selection for increased leaf area had driven down the overall population mean leaf area. This was our first "real world" experience with the problem that Griffing (1967) had addressed by recommending group selection as a way to guarantee a positive response to selection. There may be no real difference between a "herd of fleet deer" and "fleet herd of deer," but there is a large difference in total yield between "a crop of high-yielding plants" and "a high-yielding crop of plants."

The among-deme variation in shading reduced the response to group selection only in those metapopulations without individual selection. Just like Craig (1982), Charles discovered an interaction between individual and group selection: the response to group selection was strongest in the treatments without individual selection. What made this finding surprising was that individual selection for increased leaf area interfered with group selection for increased leaf area so the total response was weaker even when both levels of selection were acting in the *same direction*! The interference effect was strongest in the constant environment, where individual selection was also strongest. Charles suggested that individual selection interfered with group selection by using up some of the genetic variation that might otherwise be partitioned among populations by random drift.

These results showed conclusively that, for some traits, group selection could succeed where individual selection could not. Indeed, the scale of environmental variation could affect and even reverse the relative efficacy of the two levels of selection. Furthermore, we suspected that this result might be common for traits involved in competition for scarce resources, a concept central to Darwin's theory of natural selection.

Diversifying My Research

Around this time, in response to graduate student demand, Steve Arnold and I began co-teaching a graduate seminar on sexual selection. The typical introduction to sexual selection emphasized a necessary relationship first noted by R. A. Fisher: the average number of offspring per female had to equal the average per male, because every offspring has one mother and one father. A presentation in that graduate seminar by my student, I. L. Heisler, made me realize that I might be able to partition variation in male fitness into two separate components, one within harems and one among them, in the same way I had partitioned fitness variation into separate components within and among groups. By viewing harems as "groups of females," I discovered a necessary relationship between the sexes involving the variance in fitness (Wade 1979b) that paralleled Fisher's relationship between mean fitnesses of the sexes. My model showed that selection on males was necessarily stronger than that on females whenever some males had many mates while others had none (Wade 1979b). After reading my manuscript, Lande pointed out that J. F. Crow (1958, 1962) had partitioned the variation in human relative fitness into separate components of viability and offspring number, calling the variance in relative fitness the intensity or opportunity for selection. For that reason, I called the variation in relative fitness associated with the variance in mate numbers the intensity or opportunity for sexual selection. Our seminar on sexual selection was the first of many times that an insight from my teaching led directly to a publication.

These and other concepts combined with the elegant quantitative genetic models of Lande and Arnold formed what came to be called the *Chicago School of Evolutionary Biology*. My postdoc Dr. C. R. Boake brought an exceptional background in animal behavior from Cornell to the Wade lab as well as a keen interest in sexual selection (Boake and Capranica 1982). In *Quantitative Genetic Studies of Behavioral Evolution* (Boake 1994), she summarized and expanded the influence of this diverse body of "Chicago-style" research.

Conclusion

Taken together, these studies with *Tribolium* and *Arabidopsis* showed that the efficacy and the domain of group selection were much broader than the narrow parameters under discussion in the group selection debates. Traits that did not respond to individual selection could and did respond to group selection. Although environmental variation, like competitors, fluctuations in N, or shading, lowered g^2, such variation did not prevent a response to group selection. Furthermore, we had observed correlated responses in other population traits not directly targeted by group selection. Many of the correlated population responses to group selection involved the most frequently discussed topics in ecology—namely, competitive ability, population outbreaks, and oscillatory density dynamics.

Lastly, the relaxation of group selection did not lead to a loss of selected gains. On the contrary, the high and low groups maintained their characteristics. Moreover, the process of genetic differentiation continued after group selection was relaxed and it involved an interaction between drift and individual selection. Given a strong response to group selection, its persistence in the face of relaxed selection, and the lack of response to individual selection, the evidence was mounting that there were components of genetic variation available to group selection that were not available to individual selection. Population heritability, g^2, was composed of different genetic elements than individual heritability, h^2. The different multilevel adaptive processes appeared to have distinct genetic domains.

Despite these results, our findings appeared to have little or no impact on the ongoing group selection debate within evolutionary biology, although they did have impact in History and Philosophy of Science (e.g., Wimsatt 1980; Lloyd 1984, 1988). Within evolutionary genetics, the focus of discussion (until very recently) remained the "additive" genetic variance. Our enthusiastic discussions of gene interactions (epistasis) and genotype interactions (indirect genetic effects) served only to invite additional criticism from those invested in the primacy of evolution as a one-gene-at-a-time process where a gene's evolutionary fate depended solely on its intrinsic, additive effect on fitness.

7 The Evolution of Sociality

The Blackstone Institute

Sociality on the south side of Chicago was enriched during the late 1970s. The tradition of an evolutionary biology Halloween party began at a graduate-student apartment on Blackstone Avenue. It featured elaborate costumes with themes ranging from biology (pterodactyls and ctenophores) to popular culture (Sid Vicious' girlfriend and *The Toolbox Murders*) to philosophy and religion (pasta farians). Each party culminated with the McCauley-inspired "pumpkin suicide," off the third floor balcony.

Musical tastes spanned an arc from Fleetwood Mac, to Thelma Houston, the Talking Heads, Prince, and, eventually, to Lori Anderson. A favorite was the reggae anthem to tenure and promotion, *You Can Get It If You Really Want*. The morning Chicago talk-radio show, "Steve & Garry's Breakfast Club," segued, at the instigation of Felix Breden, into "Post-doc Breakfast Club," which remains a Wade-lab institution. These and other party activities (e.g., Horrible Food Potluck Dinners, Fall Harvest Parties, Fieldtrip Breakfasts) were our antidote for the mind-numbing tedium of *Tribolium* research.

Maynard Smith's Kin and Group Selection

Hamilton's seminal paper on kin selection and the evolution of altruism (1963, p. 355) cited only the founders of evolutionary genetics: Wright, Fisher, and Haldane. There, he introduced his famous rule governing social evolution: "If the gain to a relative of degree r is k-times the loss to the altruist, the criterion for positive selection of the causative gene is $k > 1/r$. Thus a gene causing altruistic behavior towards brothers and sisters will be selected only if the behavior and the circumstances are generally such that the gain is more than twice the loss; for half-brothers it must be more than four times the loss; and so on." Hamilton identified his r as *Sewall Wright's Coefficient of Relationship*. Instead of r, Wright (1943, 1945, 1969) used F_{ST}, the fraction of the genetic variation among groups, which he variously called the coefficient of relationship, the inbreeding coefficient, or the fixation index depending on the topic.

Maynard Smith (1964, pp. 1145, 1147), building on the work of Hamilton (1963, 1964 a and b), distinguished kin selection from group selection in this way: kin selection could occur with random mating, while group selection required inbreeding (mating between genetic relatives). Despite its dependence upon r, kin selection did not require inbreeding. Although Wright had used F_{ST} as a measure of inbreeding and recognized that continued random drift leads to high r, Maynard Smith drew a sharp distinction between consanguineous mating (inbreeding sensu strictu) and other means of generating genetic relatedness.

Because kin selection explained social behavior of individuals, many considered it individual selection (West-Eberhard 1975). However, from the process perspective, I could not distinguish the "individualistic kin selection" of animal behaviorists from the family selection of animal breeders. Breeders recognized family selection as group selection, employed it in several circumstances (chapter 2), and recognized that its efficacy was enhanced by inbreeding. Breeders used and measured the efficacy of group selection every day, especially in regard to yield in crop plants. In contrast, at a Chicago seminar in 1975 or 1976, Dr. R. Trivers stated that the predictions of kin selection theory could not be tested in the laboratory; they could *only* be tested with taxonomic comparisons between species with different values of r (e.g., Trivers and Hare 1976; see critique in Lloyd 1988/1994 reissued by Princeton University Press).

I took up the challenge to design an experimental study of kin selection using *Tribolium*. Kin selection theory was an explanation for positive as well as negative social behaviors, like infanticide and cannibalism (e.g., Hamilton 1970; Chapman and Hausfater 1979; Sherman 1981). Our beetles had an essential prerequisite for a selection experiment: heritable variation in cannibalism. For genetic reasons, the beetles of some populations were more prone toward cannibalism than those from other populations.

Following Park et al. (1974), I manipulated genetic relatedness between potential larval cannibals and their egg victims (Figure 7.1), creating three distinct levels of r (½, ¼, or 0) between cannibal and victim. Cannibalism itself generated the force for group selection. All groups were founded with the same number of larvae and eggs but variations in cannibalistic behavior resulted in variations in group size: groups with little cannibalism were larger than those with more cannibalism. Because larger groups contributed more offspring to the next generation than smaller groups, group selection favored reduced cannibalism.

In establishing the next generation, I varied the genetic correlation between mates *independently* of the genetic correlation between cannibal and victim. That is, for each value of r between larvae and eggs, I could impose within-group mating (inbreeding) in one treatment but random mating in another. This allowed me to directly address Maynard Smith's claim that inbreeding distinguished kin from group selection and his claim that kin selection was a more potent evolutionary force than group selection. If he were correct, the value of r between cannibal and victim would determine how cannibalism evolved in response to kin selection (random mating) but, with group selection (inbreeding), there would be little or no change in cannibalism. On the other hand, if kin selection was a form of group selection, I predicted that cannibalism would evolve more rapidly with inbreeding, because inbreeding would increase the group heritability.

One of the nicest features of this experiment was that none of the selection was imposed by me. Larvae created the selection themselves by eating or by not eating eggs (see schematic diagram, Figure 7.1). With an r of ½, the surviving eggs shared genes with their larval cannibals. With an r of 0, the surviving eggs were not genetically like the cannibals. Cannibalism should decrease when r was ½ because the many surviving eggs of the least cannibalistic larvae shared genes with the reluctant cannibals. Cannibalism should not change when r was 0, because, although the larger groups are the result of more eggs surviving, those eggs have nothing genetically in common with the larvae that spared them.

Selection occurred as the larger groups made a greater genetic contribution to the next generation than the smaller groups. This group selection occurred regardless of the mating system. *Why should it matter whether survivors mated with their siblings or mated at random?* It mattered because in-

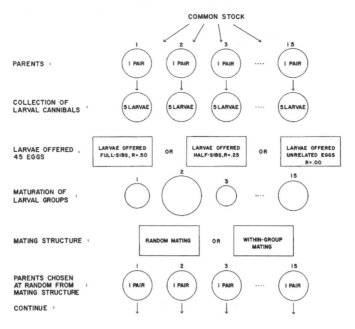

PARENTS :

COLLECTION OF LARVAL CANNIBALS :

LARVAE OFFERED : 45 EGGS

MATURATION OF LARVAL GROUPS :

MATING STRUCTURE :

PARENTS CHOSEN AT RANDOM FROM MATING STRUCTURE :

CONTINUE :

FIGURE 7.1 This is a schematic diagram of the kin and group selection experiment (from Wade 1980a). The boxes mark our manipulations of relatedness between cannibal and victim or between mating males and females. The differences in group size at larval maturation are indicated by the size of the circles. These differences were caused by the larvae eating lots of eggs (smaller circles) or refraining from eating eggs (larger circles).

breeding *increased* the genetic variation among groups, thereby increasing g^2, while random mating *decreased* it. Differently put, g^2(inbreeding) was greater than g^2(random mating). Random mating blended together cannibalism genes from large and small groups, reducing g^2 in the same way that migration had reduced g^2 in our earlier experiments.

Results

As I predicted but contrary to what Maynard Smith predicted, with random mating, the rate of cannibalism did not change despite the variations in r between larval cannibals and egg victims. With inbreeding, however, cannibalism declined whenever the cannibals and victims were genetically related. Moreover, it declined more when r was ½ than when r was ¼. Consonant with previous results, the response to the group selection was larger when g^2 was higher. Our findings supported the inference that kin and group selection were the same process and that the g^2 resulting from drift and the mating system determined the efficacy of the process. If one adopted the distinction between kin and group selection proposed by Maynard Smith (1964, 1976), then our results showed that group selection was more effective than kin selection.

Our most important finding was that social context and mating system determined whether cannibalism increased or decreased. A single population with genetic variation for cannibalism had evolved in different directions when placed in different metapopulations! Moreover, in *Tribolium*, cannibalism was a complex trait, because egg encounter rates were affected by the rate of larval development, the rate of tunneling through flour, and mandible size. If population structure influenced the evolution of cannibalism, it must then influence the evolution of each of these other traits affecting the expression of cannibalism. None of these individual traits considered alone would provoke a discussion of group selection. Yet, because each affected the expression of a social trait, their evolution necessarily involved group selection.

From Experiment to Theory

Mixing of larvae and eggs from different families in the kin selection experiment led me to extend my one-family altruism model (Wade 1978b, 1979b) to groups of two, three, or more families (1982b, 1984b). I also wondered whether total change in altruist gene frequency, Δp, could be separated algebraically into two components, one for within-family selection and the other for between-family selection. Once I had assigned to each family a fitness value relative to other families, such a partitioning was simple and the total change in gene frequency, Δp, did prove to be the sum of two parts (Wade 1980d). The first part was the average effect of individual selection within families, $\Delta p(\text{individual})$, on the altruism allele frequency. This part was always negative because only the altruists bore the fitness cost of their behavior and consequently experienced a relative fitness disadvantage. The second part was change owing to among-family selection, $\Delta p(\text{family})$. It was always positive because families with a higher frequency of altruistic members had a higher fitness than other families. The two levels of selection were acting in opposite directions. Which would prevail and under what conditions? When was $\Delta p(\text{family})$, favoring the gene, large enough to overcome opposing $\Delta p(\text{individual})$?

Selection among families triumphed over individual selection and altruism evolved whenever Hamilton's Rule was met (Wade 1980d)! When it was not met, individual selection predominated and genetic altruism did not evolve. The results were the same as in Wade (1978b, 1979b), but partitioning gene frequency change into its causal, opposing components was sufficiently novel for it to be published in *Science* (Wade 1980d). The model showed that Hamilton's theory was not "individualistic kin selection," but rather a heuristic device for determining the relative strengths of two opposing levels of selection.

As I extended the model to larger groupings of k families (Wade 1982b, 1984b), I found, in every case, that individual selection within groups worked against the altruism gene ($\Delta p[\text{individual}] < 0$); selection between groups favored it ($\Delta p[\text{group}] > 0$); and total selection was the sum of these two opposing components ($\Delta p[\text{total}] = \Delta p[\text{individual}] + \Delta p[\text{group}]$). Moreover, in all cases, when Hamilton's Rule was met, group selection was stronger than individual selection so that $\Delta p(\text{total})$ was positive. What greater irony could there be for evolutionary theory: just as Williams (1966) suppressed all discussion of group selection, Hamilton (1963, 1964) and Maynard Smith (1964) reintroduced it under another name.

Multifamily Groups

Extending the model further (Wade 1982b, 1984b), I let some groups consist of a single family, while other groups consisted of many families, analogous to the variation in N experiments that we were conducting with beetles. No matter what the distribution of families per group, I found that $(1/2k)$, the coefficient of relatedness within groups, was simply replaced by $(1/2H)$, where H was the harmonic mean group size, a finding that Wright (1938) had discovered decades earlier.

Extending the theory allowed me to make a richer set of predictions, which could be tested with experiments and field observations. These models gave me a tool for understanding the dynamics of social evolution — that is, which biological features of a species sped it up and which slowed it down. This allowed insights into the variety of social behaviors evolved by the bees, ants, and wasps as well as the reasons for the lack of sociality among other organisms. (This topic is now known as "skew theory," where the work above is rarely cited [cf. the recent review by Nonacs and Hager 2011]. This is in large part because my emphasis on group selection does not conform to the "individualistic" explanations for social behavior based on inclusive fitness [see chapter 8].)

Haplodiploidy and Altruism

I extended the model to haplodiploids, where males have only one copy of each gene but females have two (Wade 1982b). This was the system of inheritance of the bees, ants, and wasps. In kin selection theory, the altruistic sterility of the workers was explained by relatedness. Female workers were willing to forgo reproduction because they were more closely related to their sisters ($r = \frac{3}{4}$) than they were to their own offspring ($r = \frac{1}{2}$). My results were different from those of Hamilton and kin selection theory. The condition for the evolution of altruism in my model was the same in haplodiploids as it was for diploid families (Wade 1982b).

However, in changing the model, I had changed only the system of inheritance, while Hamilton had changed two things, the system of inheritance *and* the system of altruism. In Hamilton's haplodiploid model, *only* diploid females were altruists, not their haploid brothers. In my models, both sexes were altruists. When I changed the sex ratio from an equal number of males and females to one biased toward daughters, I found that altruism was easier to evolve within a haplodiploid species. With only females helping females, I obtained Hamilton's result. This finding meant that haplodiploidy alone was not sufficient to make the evolution of sociality more likely with this type of inheritance. It also required biasing the performance of the altruistic behavior and its fitness benefits toward diploid females and away from haploid males, not the most parsimonious of preconditions for the evolution of eusociality.

Over the next several years, Felix Breden, my graduate student (and later postdoc), and I developed a more complete and formal theory of the evolution of social behaviors in complex groups. Our mathematical investigations paralleled our experimental work on population fitness.

Are Social Systems Vulnerable to Cheaters?

Discussions of the evolution of altruism often contain references attesting to the vulnerability of social systems to invasion by mutations for cheating or selfish behavior. One of the earliest such discussions is found in Haldane (1932), who showed that, when a population consisted of small groups of genetic relatives, a gene for a "socially valuable but individually disadvantageous" character could increase and spread to fixation by selection (Haldane 1932, p. 207). However, in discussing his findings, Haldane (1932, p. 210) asserted that, in all such populations, "the reverse mutation may occur, and is likely to spread." He concluded that, since any system of altruism is invasible by selfish mutations, "it [is] difficult to suppose that many genes for absolute altruism are common in man."

Maynard Smith (1964) also believed that the evolution of altruistic behavior by group selection was hard to explain because a population would always be invasible by selfish or cheating mutations: "'anti-social' mutations will occur, and any plausible model of group selection must explain why they do not spread." Dawkins (1976, pp. 126, 198) too averred that "any altruistic system is inherently unstable because it is open to abuse by selfish individuals, ready to exploit it. . . . Cheats do better than indiscriminate altruists because they gain the benefits without paying the costs." Barash (1979, p. 162, 163), in a similar vein, emphasized that "group selection founders on the shoals of selfishness. It is true that groups containing altruists might do better than groups without them, but individual altruists within those groups would lose out to their selfish colleagues. . . . Individuals who place the group above themselves will be at a disadvantage compared to others who selfishly insist on 'doing their own thing.'" More recently, West et al. (2006, p. R482) summarized the vulnerability of society to exploitation by cheaters in this way: "we would not expect altruistic behaviours to be maintained in a population—put formally, altruism should not be evolutionarily stable."

One way around this problem is to argue that there are no natural examples of strong altruism (altruism directed solely toward others with no direct benefit to the altruist) because there are no relatives for the first mutant altruist to help (e.g., Santos and Szathmáry 2008). Clearly, this argument does not apply to mutants with negative effects on the fitness of conspecifics, such as the first cannibalistic mutant. The origin problem for strong altruism is similar to that of warning coloration, conspicuous coloration designed to advertise an individual's noxiousness or toxicity to predators. Once predators learn to associate color and noxiousness, warningly colored individuals enjoy higher viability owing to reduced predation. The problem for the first mutant is two-fold. First, it is more conspicuous to predators. Second, it does not gain in viability because it is a novelty to predators. Thus, there is an initial barrier to its spread (Mallet and Joron 1999). Similar barriers to initial spread occur with several other genetic elements—for example, cytoplasmic sterility caused by the microbe *Wolbachia* (Stevens and Wade 1990) or the maternal effect selfish gene, *MEDEA*, in flour beetles (Wade and Beeman 1994). Just as there is in each of these other cases, there is a variety of methods that mitigate the argument that the evolution of strong altruism is impossible.

The perception that sociality is vulnerable to cheater mutations makes little sense from the perspective of theoretical evolutionary genetics. There it is routine to add mutation to models of selection to determine the equilibrium between mutation and selection and thereby predict the standing level of genetic polymorphism within a population when selection and mutation are in balance. The repeated origin of deleterious genes owing to mutation is not a devastating threat to adaptive evolution by natural selection. For a gene that increases the fitness of an individual, its frequency change, Δp(selection), is positive. It spreads throughout a population from rare, when it first occurs by mutation, to abundant by natural selection. Mutation opposes this process by continually introducing deleterious variants of that gene into the population. Mistakes made by the DNA replication machinery during the production of gametes inevitably create poorer, defective, mutant copies. Hence, mutation changes allele frequency in the opposite direction to natural selection; Δp(mutation) is negative. Yet, mutation does not win out over Darwin's adaptive process. Instead, the opposing forces of natural selection and mutation achieve an evolutionary equilibrium. At that equilibrium, for every copy of a bad gene pushed out of a population by natural selection, a new defective copy is reintroduced by mutation. This equilibrium is called the *mutation-selection balance* and q^*, the frequency of the "bad" gene at this equilibrium, is determined by the relative strengths of selection and mutation. If s is the selection coefficient, representing the increase in individual fitness for each copy of a "good" gene, and μ is the frequency of mutational change from the "good" gene to it deleterious form, then the equilibrium frequency of the deleterious allele, q^*, is the ratio (μ/s). Since μ takes values in the range of 1×10^{-6} to 1×10^{-9}, while s is believed to be nearer 1×10^{-1} to 1×10^{-4}, natural selection has a stronger voice in the equilibrium than does mutation. As a result, q^* tends to be very small (less than 0.001). Although most mutations are deleterious and are much more numerous than favorable mutations, mutation is no threat *at all* to adaptive evolution by natural selection. Why should it be so different for a social behavior? Why should mutation be a fearfully powerful, all-destructive force when opposing kin selection but not when opposing natural selection?

Consider the "non-altruistic" allele that the altruistic allele replaced during its spread from rare to common. This non-altruistic allele is a "selfish" gene, because the individuals that harbor it do not perform the altruistic behavior, yet they reap the fitness benefits from those that do. Those with the non-altruistic allele are often called "cheaters"; they benefit from altruists but bear none of the costs. When can a population consisting *entirely* of cheaters be invaded and completely transformed into an altruistic population? Hamilton's Rule holds the answer. And, for exactly the same reason, the process that favors the spread of the altruistic allele cannot be rerun backwards in time simply by the introduction of a selfish mutation into a population consisting *entirely* of altruists.

A mutation is no more a threat to a social system evolved by group selection, than it is to an "ordinary" adaptation evolved by individual selection. To see why this is so, consider a population divided into groups. Imagine further that an altruistic gene, A, with fitness costs and benefits that satisfy Hamilton's Rule, arises and spreads through this population. Let mutation introduce a single mutant copy, *a*, of the altruism gene into this entirely altruistic population. A heterozygous individual carrying this mutation does not behave as altruistically as its homozygous group members, but it does reap the fitness benefits from them. Because this selfish (or at least less altruistic) individual does not incur the fitness cost of altruism, it is at a selective advantage relative to altruists within its own group. Does this selfish gene spread throughout the population, destroying the social system, as Haldane, Maynard Smith, Dawkins, Barash, West-Eberhard, and others feared?

No, it does not. Felix and I showed that it spreads if, and only if, $(1/2k)$ $b_{Total} < c$ (Wade and Breden 1981; Breden and Wade 1981). This is the *opposite* of Hamilton's Rule, the condition necessary for the existence of the altruistic system in the first place. Could this equation be stood on its head simply by the advent of a selfish mutation somewhere in the population? The fitness costs and benefits are defined by what the altruist genotypes do, not by what the genetically selfish individuals do—they do nothing to incur a fitness cost, but simply reap the fitness benefits. So, the fitness effects do not change. The habit of living in genetically related groups does not change by the advent of a single mutation in one such group, so *r* remains unchanged. Because Hamilton's condition for the *spread* of altruism is not changed by adding one or more cheaters, a selfish mutation does not spread through the population and is no threat at all to its social system. Hamilton's Rule gives the conditions for a single altruist to invade and replace a population of "original" cheaters. As long as it holds, mutant cheaters share the fate of the original cheaters—they are replaced by altruists.

One can go further and show, for any social system evolved by kin selection, there is a *kin selection–mutation* balance: an evolutionary equilibrium between group selection favoring an altruistic allele and mutation to non-altruistic or selfish mutant copies. This equilibrium has the same properties as the standard mutation-selection equilibrium. It occurs when for every copy of a "cheating allele" pushed out of the population by group selection, another is introduced by mutation. Cheating or selfish mutations will *always* be present in a population at the kin selection–mutation balance, but they will remain rare, at this equilibrium level (Van Dyken, Linksvayer, and Wade 2011). As a result, social systems will always bear a "load of cheaters" just as ordinary adaptations bear a "load of mutations" (Muller 1950). These cheaters are no more a threat to an adaptive social system than ordinary mutations are to adaptation in general.

How frequent will cheaters be at the evolutionary equilibrium of the *kin selection–mutation* balance? As in standard evolutionary theory of selection-mutation balance (e.g., Charlesworth and Charlesworth 2010, p. 160), the answer depends upon the strength of group selection. The stronger group selection is, the lower will be the frequency of cheaters at the kin-selection mutation balance. The weaker it is, the more cheaters there will be at equilibrium.

The perception that cheaters are a relentless threat to complex social systems is a fallacy. It persists because of the bias toward anthropogenic, individual-centered logic in discussions of social evolution. The cheater fallacy is a consequence of the privileging of individual selection over other evolutionary forces. Uncritical arguments implicitly grant greater weight to "within-group" individual selection than to group selection, even when discussing models that explicitly state that the opposite must be the case for the sociality to evolve in the first place. If Hamilton's Rule were correctly grasped as the condition necessary for group selection to outweigh opposing individual selection (as Hamilton himself [1975] recognized), the cheater fallacy would have been laid to rest long ago.

Beyond Two Levels of Selection

Many organisms have population structures more complex than family groups with several hierarchical levels of organization. Populations of amphibians are a good example. Multiple ponds can be found in the same geographic region. Within ponds, tadpoles tend to aggregate with their siblings and proximate groups merge together into larger schools. The population structure thus resembles a nested hierarchy with four levels: local ponds, aggregations within ponds, families within aggregations, and individuals

TABLE 7.1 The mathematical conditions for individual, family, group, and total selection to favor the gene (positive) or oppose the gene (negative) based on fitness effects of the genetic behavior at the individual (α), family (β), or group (γ) level.

Component of Selection	Selection is Positive	Selection is Negative
Total Selection	$(\beta/4) + \alpha > 0$	$(\beta/4) + \alpha < 0$
Individual Selection	$\alpha > 0$	$\alpha < 0$
Family Selection	$(\beta - \gamma)/2 - \alpha > 0$	$(\beta - \gamma)/2 - \alpha < 0$
Group Selection	$(\beta + \gamma)/2 + \alpha > 0$	$(\beta + \gamma)/2 + \alpha < 0$

within families. Mosquitos are similar: groups within local pools, aggregations within groups, families within aggregations, and individuals within families. Much like cannibalism in beetles, individual mosquito larvae secrete waste products into the environment—products that have a negative impact on the fitnesses of other larvae. Older or larger larvae can be less sensitive or less vulnerable to waste than younger or smaller ones. Thus, the behavior of each individual has effects on its own fitness as well as on those of its neighbors, whether immediate kin or not. And, one neighbor may not be interchangeable with another. Members of one family may experience different fitness effects than members of other families. I decided to extend the population genetic models from the two opposing levels of individual and group selection, to three levels, by adding a second level of group selection (Wade 1982b).

I imagined a situation in which an individual's behavior has 3 kinds of fitness effects: (1) an effect on its own fitness, α; (2) an effect on the fitness of each of its family members, β; and (3) an effect on the fitness of each member of the other family, γ, in a two-family group. This was a model with 3 levels of selection, selection between individuals within families, selection between families within groups, and selection between groups of families (Wade 1982b). By choosing values for the three fitness effects appropriately, I could cause each level of selection to work in concert with or in opposition to any other level (Table 7.1).

When an individual treats everyone in its own and the other family in the same way (so that β is the same as γ), then there is no selection between families within groups and this three-level model reduces to the earlier two-level model of kin selection where k is 2 (i.e., two-family groups; Wade 1984b).

Within groups, Hamilton's Rule still describes the necessary condition for family selection to be stronger than individual selection. However, that criterion alone does not determine whether the behavior will evolve or not. There is the third level of selection to consider before one can determine whether or not total selection favors evolution of the behavior. One could derive a rule for group selection to override opposing family selection, but the behavior still might not evolve because of opposition from the third and lower level, selection among individuals. I found conditions where the highest level of selection could overcome *both* of the lower levels of selection working in concert against it (Wade 1982b). Adding a higher level of selection changed everything, just as adding a lower level of gametic selection changes the outcome of individual selection in the standard model (chapter 1).

Inbreeding and Kin Selection

By the early 1980s, some authors were suggesting that kin selection was a type of group selection (e.g., Bell 1978; Grant 1978; D. S. Wilson 1979, 1983 a and b; Michod 1980; Uyenoyama 1979). But, this suggestion was very strongly opposed despite the accumulating theoretical and experimental evidence (Wade 1980 a and d), which included a field study of group selection in burying beetles (D. S. Wilson 1983b). Dawkins (1979, 1982), who favored an extreme version of the gene's eye view of the evolutionary process, had listed the idea that "kin selection is a form of group selection" as one of twelve "common errors" of misunderstanding in discussions of sociobiology. And, Maynard Smith (1982) reiterated his 1964 claim that "mating within groups" was a key feature distinguishing kin selection from true group selection. In the lab, we discussed how small those deviations from random mating had to be in order for a process to qualify as group selection.

In the cannibalism experiment, I had explored complete random mating or complete inbreeding, the extreme ends of a continuum of possible mating systems. Most natural populations, including humans (Bittles and Black 2009), are somewhere in the middle with a mixture of random mating and inbreeding (Bell 1978). We wondered whether a small amount of inbreeding might have an unusually large effect on the rate of social evolution. Small values of F_{ST} in our beetle research had had large effects on g^2 and, thus, on the efficacy of group selection. We needed to see if this was true for small amounts of inbreeding.

Felix and I extended our family selection models to include arbitrary combinations of inbreeding and random mating (Wade and Breden 1981;

Breden and Wade 1981) by varying the proportion of the population that mated with siblings from 0 to 1, with the remainder mating randomly. Importantly, our model was like the cannibalism experiment: inbreeding did not affect the fitness costs or benefits. It changed *only* the frequencies of the different kinds of families and, in doing so, changed the family heritability. Because inbreeding increased the genetic differences between groups at the same time that it made individuals within them more genetically similar, we expected that it would increase the efficacy of group selection *at the expense of* individual selection. What we did not know was the magnitude of this effect. We quickly discovered that there was no analytic way to combine inbreeding with selection (Wade and Breden 1981, equations 6 through 9). We had rediscovered R. A. Fisher's model building dictum, "Either inbreeding or selection, never both at the same time." Without selection, we could see that *any* amount of inbreeding increased the frequency of the two most extreme family types, those with all altruists and those with none. These were the "armadillo" families—there was no genetic variation in either of these two families, so individual selection could not act within them, even if we had had selection in the model. We inferred that inbreeding would diminish the strength of individual selection by increasing the relative contribution of these two families to its average.

Felix devised a simulation allowing us to explore a variety of different fitness costs and benefits as well as different levels of inbreeding and random mating. We discovered that even small amounts of inbreeding greatly increased the rate of evolution of altruism. For example, the rate of social evolution in a population with 20% inbreeding (and 80% random mating) was *fifty-fold greater* than that in a population with no inbreeding at all (100% random mating). With exactly the same costs and benefits, social evolution in a population with 100% inbreeding was *five hundred times faster* than in a population with 100% randomly mating. It was no wonder that my experimental studies had revealed such a strong and critical effect of the mating system on the evolution of cannibalism!

Inbreeding affected the *response* to selection via its effects on group and individual heritability. It increased the between-group genetic variance and decreased the within-group variance, enhancing the efficacy of group selection and diminishing that of individual selection. Introducing Maynard Smith's "discontinuities in breeding structure" into our model systematically favored group selection at the expense of individual selection. And, even slight discontinuities were a tremendous accelerant of social evolution. If "discontinuities in breeding structure" were the salient feature distinguishing group from kin selection, these discontinuities also made group selection much more effective than individual selection.

Mating and Associating with Kin

In the kin selection experiment (Wade 1980a), I had manipulated the breeding system (inbreeding or random mating) and the "association system" (the relatedness between interacting cannibals and victims). We realized that, with a slight modification of our model, we could investigate the effects of mating with relatives separately from the effects of associating with relatives, just as I had done with the beetles. We could investigate whether mating with relatives was more important than associating with relatives, an inference I had drawn from the cannibalism study.

We simulated two-family groups of females mating and producing offspring in proximity to one another (Wade and Breden 1987). We varied the mating system by varying the fraction of females mating randomly with their siblings, randomly with members of their own or the other family, or randomly across the entire population. We also varied how the families associated by varying the fraction of females that associated with a sibling, with another female drawn at random from her two-family group, or with a female randomly drawn from the population. At one extreme, a female could mate with a sibling and associate with a sibling. At the other, a female could mate at random and associate with another random female. In between, a female could mate with a sibling but associate with a random female or she could mate at random but associate with a sibling. We could watch the evolution of altruism in each of these complex social groups and determine unequivocally whether mating with sibs or associating with sibs had the greater effect on social evolution.

Our findings are summarized in Table 7.2. Each cell in the table gives the number of generations necessary for an initially rare altruism gene to spread to fixation in a specific type of population. For every population, the behavior had the same fitness costs and benefits; they differ from one another in how the females mated and in how they associated with one another. For example, consider the first row of Table 7.2. For all populations in this row, females associated with siblings to form two-family groups, but they chose their mates in three different ways. Altruism evolves ten times faster if females mate with siblings (30 generations to fixation) than if they choose a mate at random from the general population (300 generations to fixation). Similarly, consider the first column. Here, females always mate

TABLE 7.2 The number of generations required for an altruistic gene to go from an initial frequency of 0.05 to a final frequency of 0.95. The row averages illustrate the effect on the rate of evolution caused by associating with individuals of different genetic relatedness. The column averages illustrate the rate of evolution caused by mating with individuals of different degrees of relatedness (Wade and Breden 1987).

Family Association	Mating Association			
	Within Family	Within Group	Within Population	*Row Average*
Within Family	30	35	300	*120*
Within Group	30	35	370	*145*
Within Population	70	225	520	*272*
Column Average	*43*	*98*	*397*	

with siblings, but they associate in different ways to rear their young. Altruism evolves faster if they associate with siblings (30 generations to fixation) than if they associate with a female at random from the general population (70 generations to fixation).

The *row averages* (far right column of Table 7.2) indicate how fast altruism evolves for a given way of forming two-family groups, no matter how females select mates. The *column averages* (bottom row of Table 7.2) tell us how fast altruism evolves for a given way of mating, no matter how females associate into two-family groups. The rate of social evolution varies ten-fold across the column averages but only two-fold across the row averages. This means that a female's genetic relatedness to her mating partner is a much more important determinant of the rate of social evolution than her relatedness to her associates.

It would be possible, as Maynard Smith (1964) argued, for association groups to be "virtual," existing only in the minds of altruists, as long as altruists could discriminate kin from non-kin and alter their behavior accordingly. Our results showed that associating with relatives does increase the rate of social evolution. Nevertheless, our results also showed that the choice of mating partner had a much stronger effect than decisions to discriminate for kin and against non-kin. The results of experiment (Wade 1980a) and theory (Wade and Breden 1987) were concordant and reinforcing. They showed that it made little sense to discriminate kin selection from group selection on the basis of inbreeding; they were the same evolutionary process. By partitioning the separate contributions of individual and group selection to the evolution of sociality, it was easy to see that inbreeding enhanced the efficacy of group selection favoring altruism at the same time that it diminished the efficacy of individual selection opposing altruism. That is why mating system was more important to social evolution than the kin recognition system.

Synergistic Coevolution of Sociality and Mating System

Felix thought that some features of social evolution might mirror those of Fisher's model of sexual selection. Fisher (1958) had proposed a model of "run-away" sexual selection. Like Darwin, Fisher considered the exaggeration of some male traits, especially plumage in birds, to have evolved by sexual selection driven by female mate choice. That is, males with the most exaggerated plumage were preferred as mates by females, and such males had many mates, and thus, many offspring, as a consequence. Conversely, less ornamented males were rejected by females, and often did not get to mate at all. Fisher further imagined that females varied from one another in their "choosiness," with less choosy females willing to accept almost any male as a mate but the choosiest accepting only the most ornamented males. This addition to Darwin's idea had two consequences. First, the most ornamented males not only obtained more mates but also their mates were "choosier" on average than the mates (if any) of less ornamented males. As a result, the most ornamented males had more extremely ornamented sons as well as choosier daughters. In this way, over time, females become more and more extreme in their mating preferences, driving the evolution of ever more extreme male ornamentation. This self-accelerating process is called *run-away sexual selection*.

Felix suspected that something similar might occur in social evolution. Families with a high proportion of altruists have higher fitness than families with fewer altruists. If offspring from these families also exhibited a propensity to inbreed, then, in the next generation, their offspring would also find themselves in families with a high level of altruism and consequently they would again enjoy higher family fitness. Just as genes for extreme female choice become coupled to genes for extreme male ornaments under Fisher's process, genes for altruism would become associated with genes for inbreeding under Felix's process. I was not completely convinced of this argument because I thought the opposite was also plausible. If genes for outbreeding became coupled with genes for selfishness and individual selection were stronger than group selection, then there might be a "run-away" *from* sociality in the opposite direction. Felix disagreed. Because a little bit of inbreeding had such a disproportionately large effect on group selection, he believed there was an inherent bias that favored run-aways toward altruism and inbreeding over those toward selfishness and outbreeding.

To explore his idea, we added a second gene to our altruism model. The frequency of this gene determined whether or not an individual mated within its sibling group or mated at random. Instead of us determining which fraction of individuals mated within families or at random across the population, the "inbreeding gene" frequency determined that fraction. We did not allow the inbreeding gene to have any effect on fitness because we wanted to see how social evolution interacted with mating system evolution. Although we could have added an element of inbreeding depression to the model (a fitness penalty for inbreeding), we did not; we were not interested in direct selection on the mating system, only in indirect selection on mating as a consequence of selection for altruism.

In our simulations of this more complex model, we calculated the population genetic parameter, D, the "linkage disequilibrium." D measures the degree of association between the altruism gene and the inbreeding gene. Positive D values mean that you tend to find the inbreeding gene in altruistic individuals. As a result, if the altruism gene increases, so too does the inbreeding gene. Negative D values have the opposite implication. We started our simulations with no association ($D = 0$); so there was no initial way for selection favoring altruism to influence the allele for inbreeding. For Felix's run-away process to work, selection had to create positive D values. But, not all positive values of D lead to run-away evolution. The rate of evolution of both genes had to be self-accelerating. That is, positive D alone would not be sufficient to trigger a run-away process in most two-gene models.

When we analyzed the simulation data, we discovered the social run-away process that Felix had predicted. The easiest way to illustrate the synergistic, run-away effect of the simultaneous evolution of altruism and inbreeding genes is to contrast it with the evolution of altruism under the two extremes, random mating and inbreeding. In Figure 7.2 below, the long-dash line is the evolution of altruism with 95% inbreeding and the solid black line is its evolution with 100% random mating. Clearly, the altruism gene increases from rare to very common much more quickly with inbreeding than it does with random mating—that we knew from the previous model (Wade and Breden 1987). The dot-dash line shows the rate of evolution when *both* genes start out rare and randomly associated. Hence, individuals who are both altruistic and prone to inbreeding are extremely rare initially: they comprise only two-tenths of one percent of the population. The vast ma-

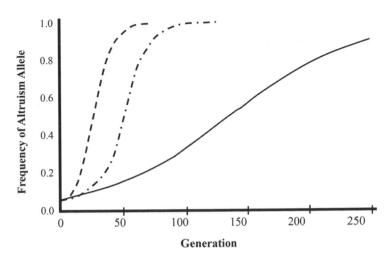

FIGURE 7.2 The change over time in the frequency of an altruistic allele is depicted for different initial frequencies of a gene for inbreeding. When the initial frequency of the inbreeding gene is 0.00 (i.e., 100% random mating) and that of the altruism gene is 0.05, the altruism gene evolves slowly (solid black line). When the initial frequency of the inbreeding gene is 0.05, however, the altruism gene evolves rapidly (dot-dash line), almost as fast as it would in a population where the initial frequency of the inbreeding gene was 0.95 (long-dash line). The altruism and inbreeding alleles become associated with one another by selection.

jority of the population mates at random and is not altruistic. Clearly, the evolutionary trajectory of the altruism allele (dot-dash line) is much closer to the trajectory of the 95% inbreeding case than it is to the 100% random mating case. This is the run-away coevolution of the sociality and mating system. Inbreeding is initially very rare, but it becomes common as the altruism gene becomes common. As inbreeding increases, so too does the rate of evolution of the altruism gene, dragging the inbreeding allele to even higher frequencies. In less than 100 generations, everyone is altruistic and everyone is inbreeding.

By monitoring the value of D each generation, we found that the gene for inbreeding quickly became positively associated with the altruism gene, even though both genes were unlinked. And, D very quickly reached a value of +0.21, close to its theoretically maximum possible value of 0.25. The evolutionary dynamics of the two genes became strongly coupled and reinforced one another, resulting in rapid increases to high frequency for both genes.

Multilevel Selection and Inclusive
Fitness as Interpretive Frameworks

The Inclusive Fitness Framework

Ever since Hamilton (1964 a and b), inclusive fitness modelers have presented social evolution from the viewpoint of the individual or the viewpoint of the gene. In this framework, an individual's inclusive fitness has two parts, one based on its viability and the offspring it produces and the other based on the offspring the individual helps others produce, weighted by its genetic relatedness to the others.

The second component is the indirect fitness effect: the offspring of others that are awarded to the altruist whose generous donation of fitness benefit makes those offspring possible. This indirect reproduction must be weighted by r, genetic relatedness, to convert the offspring of others into the fitness of the altruist. For example, if a brother produces one extra offspring owing to help from a sibling, then ½ of that extra offspring belongs to the helping sibling. If a more distant relative produces one extra offspring owing to help from the same altruist, then one r-th of that relative's extra offspring is assigned to the helping altruist. The function of r is to convert the offspring of others into "inclusive fitness equivalents" to be added to the fitness of the altruist (West-Eberhard 1975). For example, Ricklefs (1975) applied the inclusive fitness approach to explain the evolution of cooperative breeding in birds: "When we extend our concept of individual fitness to include the fitness of kinship groups, we see that because individuals are as closely related to siblings as to their own progeny, increments of parental care contribute equally to an individual's evolutionary fitness when applied to sibs or to offspring." The accounting of fitness is done differently in other areas of evolutionary biology.

In other contexts, evolutionary biologists find breeding individuals and count their offspring. They do not count the offspring of breeding individuals and then reshuffle them, weighted by r, from one individual to another. For example, in the study of plant fitness, it is routine to count the seeds on a plant and to use that count as a measure of its fitness. A given plant may well be the "father," through pollination, of seeds on other plants; but, it is often too difficult and costly to make paternity assignments and add an additional ½ offspring for each seed fertilized on another plant to the pollinating plant.

Nevertheless, when it is important to understand reproductive success through pollen and seed, highly variable genetic markers—so variable that each parent is different from every other—can be used to screen offspring, no matter where they occur, and assign each of them to the proper parent (Broyles and Wyatt 1990). This kind of "reshuffling" is not possible in principle for altruistic individuals, because assisting does not leave a genetic mark on the "extra" offspring that a relative produces. In this sense, the inclusive fitness approach is a conceptual sleight of hand. In inclusive fitness theory, offspring are taken away from the individuals that produce them and are awarded, suitably discounted, to others that have helped them. There is no empirical method to test whether Nature participates in such discounting, how often she errs, or how variable she is when genes and environments interact to affect the phenotypes of the helping or the helped.

The Multilevel Selection Framework

Multilevel selection theory takes a more direct approach to fitness accounting. Like the inclusive fitness approach, it assigns a fitness to each individual that is made of two components, one direct and one indirect. The first is the increment owing to the effects of the individual's genes on its own survival and reproduction. The second is the indirect fitness effect, the additional viability or offspring that a genotype produces because it experiences a social environment and is helped by others. There is no a priori weighting of the indirect effect by relatedness to the individual's helpers.

The variation in fitness among individuals, which determines the strength of selection, thus depends upon (1) the variation among individuals in the direct effects of their genes; and (2) the variation among individuals in the social environments they experience. The direct effects vary in proportion to the genetic variance among individuals within groups, while the indirect effects vary in proportion to the genetic variance among social environments. Relatedness appears in the expression for gene frequency change because it is the fraction of genetic variation among social groups. The weighting by r does not appear in the genotypic fitness function in the multilevel selection approach as it does in the inclusive fitness approach. Instead, in multilevel theory, r appears as the component of genetic variance in the equation describing gene frequency change, because it influences the strength of selection among social groups.

The two mathematical approaches lead to different methods for studying the effect of the social environment on individual fitness. The multilevel selection approach uses the same methods to characterize the effect of the social environment on fitness as it does to characterize the effect of any abiotic environment on fitness. Typically, individuals or genotypes from one place are transplanted to another place with a different environment. One genotype placed in two different environments may produce different numbers of offspring. Transplanting several genotypes into each of a series of environments lets one determine the effect of environmental variation on individual fitness. In environments favorable to reproduction, some, if not all, genotypes produce "extra" offspring. Because abiotic environments are not heritable, such environmental variation tends to reduce the heritability of individual traits, especially when the best genotype in one environment is not the best in another. Unlike abiotic environments, social environments can be heritable, leading to run-away evolutionary dynamics such as those we saw above between social system and mating system (Wade and Breden 1987) and for many other cases of social interaction, such as mother-offspring and sibling-sibling interactions (Drown and Wade 2014).

When placed in different social contexts, different genotypes may produce different numbers of offspring just as they do when placed in different abiotic environments. Variations in the social environment owing to the presence of altruists may allow an individual to produce "extra" offspring in some social environments but fewer in another. These extra offspring are counted as part of the fitness of their parents; they are not redistributed among the members of the social environment as they are in the inclusive fitness approach.

Consider an example. The social environment plays a major role in determining the number of eggs laid by a hen (Craig and Muir 1996; Biscarini et al. 2010). In artificial selection to increase egg lay, breeders must decide which hens to breed and which to prevent from breeding. The best hens are those with genes whose direct effects increase egg lay and those with genes whose indirect effects on the social environment are most benign and therefore conducive to egg laying by others. In evaluating the social environment, breeders do not award a hen extra "egg equivalents" based on how she treats other hens and affects their egg numbers. Instead, they infer the presence or absence of social genes in a given hen by observing what *her genetic relatives* have done to the egg production of other hens (Biscarini et al. 2010). And, to measure the effect of social environment, just like measuring the effect of any environment, breeders create variation from one social environment to another. If a hen's genetic relatives increase the egg lay of those around them by an amount s_s, then r is the probability that that an untested relative also carries genes for such a benign social environment.

For simple genetic models, there is a mathematical similarity between the inclusive fitness approach and the multilevel selection approach only in the sense that either approach can get to Hamilton's Rule. Nevertheless, they are not equivalent or alternate frameworks that can be intertranslated (Lloyd et al. 2008); the two approaches differ fundamentally. Both add a second term to individual fitness — namely, the "inclusive" term or the indirect effect, respectively, for "extra" offspring. The inclusive fitness method weights the second term of the fitness function by r, while the multilevel selection approach does not weight the indirect fitness effect by r. How can the two approaches reach even one similar conclusion if they do not agree on the specification of the underlying fitness function? Furthermore, Where and Why does r appear in multilevel selection models if it is not put into the fitness function? In multilevel selection, the value of r emerges in the Δp terms because it is essential to describing the two components of genetic *variance*, those within and between groups. The social environment contributes to heritable *differences* in individual fitness only when individuals experience genetic variation in their social environments. The r is the same, but its origins in the two approaches are very different.

The parameter r appears in the multilevel selection approach for exactly the same reason that it appears in other evolutionary genetic models of individual selection that have nothing at all to do with sociality or indirect effects. For example, inbreeding increases the rate of evolutionary change by natural selection for additively acting genes (Δp[random mating] = spq/W versus Δp[inbreeding] = spq(1 + r)/W). Yet for both random mating and inbreeding, mean fitness, W, equals (1 + 2spq). The point is that r does not appear in the fitness function, W, which is the same in both cases, but it does appear in the expression for Δp *because it describes the effect of inbreeding on genotypic variance.* The multilevel selection approach treats social and nonsocial models in the same way; the inclusive fitness method does not. The weighting of benefit by r in the inclusive fitness function is a heuristic device that obscures the underlying evolutionary process.

The individual focus of the inclusive fitness approach views individual behavior as a strategy for exploiting the social system. This perspective is the origin of the evolutionary thinking that privileges the actions of the individual over that of the group. Once assigned to individual genotypes, inclusive fitness tends to be interpreted as a property of the individual, an approach referred to as *methodological individualism*, a term from sociology referring to theories of society based on the intentions of its members (see also Cassidy 1978). The focus on individual intention is clear:

> The social behaviour of a species evolves in such a way that in each distinct behaviour-invoking situation the individual will seem to value his neighbours' fitness against his own according to the coefficients of relationship appropriate to that situation (Hamilton 1964, p. 19).

When the evolutionary dynamic is presented from the individual's strategic viewpoint instead of the underlying process, it leads to mistakes in thinking (Lloyd and Feldman 2002). It requires special, separate treatment of social environments relative to abiotic environments. Moreover, it perceives conflict everywhere because all parties in a group are separate, selfish evolutionary actors, bent on exploiting one another whenever possible until they battle to a species-specific standstill.

Cheaters Sometimes Prosper But They Never Win

I believe this individual-biased view is responsible for the persistence of the cheating fallacy. *Cheating*, as an individual strategy, is more successful than *altruism* as an individual strategy. Cheaters get something for nothing and exploit the social group for their own benefit. Therefore, societies are perpetually threatened from within by the selfish interests of their own members. Cheaters always win in the perspective fostered by the inclusive fitness framework. The logic is impeccable in large randomly mating societies, but it collapses in metapopulations with variation in social contexts from deme to deme.

The conclusion from the multilevel selection approach is the opposite: "Cheaters never win" (Wade and Breden 1981; Breden and Wade 1981; Van Dyken, Linksvayer, and Wade 2011). As with ordinary deleterious mutations, the frequency of defective social genes is determined by the mutation rate and the opposing strength of group selection. There is no need to label the individuals harboring these mutations as *cheaters*, practicing an alternative behavioral strategy.

It is possible to distinguish the inclusive fitness perspective from the multilevel selection perspective not only theoretically but also empirically. For the cheater controversy, this can be done by characterizing the frequency distribution of mutations at social genes. Multilevel selection theory predicts that this distribution will be like the frequency spectrum of other, nonsocial genes. Inclusive fitness theory assumes that rare mutant cheaters enjoy a net individual fitness advantage so that the frequency distribution of "cheater alleles" within natural populations will be different from that expected for ordinary mutations in a particular way. Namely, the frequency distribution of cheating alleles should have a "fat tail" because the selective advantage to cheaters will make copies of these genes more numerous than predicted by the group selection–mutation balance. The inclusive fitness theory makes a second prediction. The most severe mutations, those that completely knock out expression of a social gene, will be "better" cheaters than mutations that only mildly impair social gene function. A severe mutant is a better cheater because knocking out expression of a social gene reduces the fitness cost to zero, while impairing its function retains some of the fitness cost. Thus, it predicts that the most common mutations should be the most severe with respect to gene function.

What do the data say so far? First, there is no fat tail. The frequency distribution data of social genes conform to the prediction of multilevel selection theory (Van Dyken and Wade 2012). Secondly, like most ordinary genes, the large majority of mutations at social genes are of small effect rather than large effect; this is the opposite of the inclusive fitness prediction.

The continued development of sociogenomics will allow for much more incisive testing of the distinctive predictions of the two approaches (Linksvayer and Wade 2009; Van Dyken, Linksvayer, and Wade 2011). Sociogenomics will also allow us to test the unique predictions of multilevel selection theory—for example, that the frequency of defective copies of social genes ("cheaters") should vary among populations in inverse proportion to the strength of group selection.

Inbreeding: Why Small Changes in *r* Produce Large Changes in Rate

With the inclusive fitness approach, inbreeding facilitates the evolution of social behaviors because it increases *r*. This is the same reasoning that led to the suggestion that the identical quadruplet broods of nine-banded armadillos should be highly altruistic—although they are not (see chapter 3). However, the inclusive fitness approach did not predict the enormous effect that even small amounts of inbreeding have on the *rate* of evolution of social behaviors that Felix and I discovered. Moreover, from their individualistic point of view, they *could not have predicted it*. In standard theory, a 50-fold increase in Δp, the rate of evolution per generation, requires a 50-fold increase in s, the selection coefficient. Under the individualistic interpretation of inclusive fitness theory, modest levels of inbreeding produce modest increases in r; (see s_s in chapter 2). Increasing r by 20%, say from 0.50 to 0.60, would certainly relax the $rb_T > c$, Hamilton's criterion of social evolution. But, it would not increase the *product rb* 50-fold. In inclusive fitness theory, it is not possible that a small increment in the fitness weighting factor, r, could have such a large effect on the evolutionary dynamic. Nothing in that theory foreshadowed the one or two orders of magnitude increase in the rate of social evolution discovered by our multilevel approach and confirmed in our experiments (Wade 1980a).

Multilevel selection theory allows inbreeding to be understood through its effects on the heritable variation within and among groups. Inbreeding diminishes the efficacy of individual selection while, at the same time, it enhances that of group selection. Neither the direct fitness effect, c, nor the "indirect" effect, b_T, changes with inbreeding. Inbreeding does increase the variation among social environments, while decreasing the genetic variation within them. Inclusive fitness theory misses the simultaneous effects of inbreeding on reducing h^2 and increasing g^2. Whereas inclusive fitness theory is focused on mean individual fitness, multilevel selection theory is focused on the individual and group heritabilities. A poor understanding of the *process* of social evolution results in a myopic focus on the *outcome* of evolution. By misunderstanding the process responsible for the outcome, inclusive fitness makes incorrect predictions. As discussed in chapter 6, inclusive fitness theory (Maynard Smith 1964) counterfactually suggested that kin selection in randomly mating populations was stronger than group selection with inbreeding. Moreover, poor predictions make it difficult or impossible to test evolutionary genetic theory with DNA sequence data.

From Laboratory to Theory to Field

To test the predictions of our laboratory and mathematical models in Nature, we needed an abundant, primitively social insect whose ecology was easy to manipulate. While teaching Field Ecology, I found the perfect system in the Cook County Forest Preserves. Felix and I used it for over a decade to measure selection within and among groups in a natural population. These studies allowed me to calibrate our *Tribolium* research to Nature. That research effort is detailed in the following chapter.

8 Calibrating the Laboratory to Nature

Tenure and Family

The tenure process at Chicago was a mystery. Junior faculty likened it to being hired, thrown into a darkened room, and told, "There is a hurdle in here somewhere. Get over it." For the next six years, you flung yourself around in the dark until the tenure committee pounded on your door and asked, "What the hell have you been doing in there?" You then compiled your publications, and they examined the blood on the walls to see if you had made it over the hurdle.

Most of the pressure to do good work came, not from my other junior colleagues, like R. Lande and S. Arnold, or from the threat of not being tenured; rather it came from the portraits of former department members on the walls of the Lillie Room, especially W. C. Allee, A. Emerson, T. Park, R. Lewontin, and S. Wright.

In May 1980, my tenure materials consisted of the McCauley-Wade group selection studies, some of the kin selection work, and a paper on the opportunity for sexual selection (Wade 1979b). Each addressed a contentious topic and it was not clear to me what outside reviewers would think of my dossier. I applied for a position at Berkeley that year in case things did not go well at Chicago. Months of silence went by until December, when I learned that I would be a tenured associate professor beginning July 1981. In my career-long competition with Park, this meant that I had achieved tenure one year earlier than he had. Park pointed out that this could not be helped since he had been a postdoctoral fellow for a year with Dr. Raymond Pearl at Johns Hopkins University prior to joining the Chicago faculty, while I had not. He did concede that, if I remained on the faculty until age 65, I also had a chance to beat his record of 37 years as a faculty member at Chicago. As it turned out, I moved to Indiana University in midcareer, 14 years short of Park's record.

Tenure gave me the freedom to attempt riskier and longer-term research projects. The main two projects I initiated were field research on group selection (this chapter) and an experimental test of Wright's Shifting Balance Theory (chapter 9). The fieldwork was risky because it was not the sort of research the NIH would support. Although the climate of opinion appeared to be thawing toward group selection, I thought I would need two or three years of solid preliminary field data before a proposal to NSF would stand a chance of success. I searched each spring for an abundant, local organism that lived in discrete social groups while I taught graduate Field Ecology — eventually, I found one.

Tenure also gave me the nerve (and the University financial backing) to venture into the housing market in Hyde Park. National mortgage rates in 1981 were close to 20%, but, with the University's help, I was able to secure a 30-year loan at the "bargain" rate of 13.8% on a 3-bedroom cooperative apartment. My daughter Catherine was born April 14, 1983, at Chicago Lying-In Hospital. Catherine's mom, Dr. Pat McElroy, was one of the first women cardiologists of the University of Chicago hospitals and her position left little time for child-rearing, which naturally fell to me. I had already had a "practice family" in my younger brothers and sisters. There was no paternity or family leave policy at Chicago, but, because of tenure, I was able to stay home days with Catherine for her first six months, and go in to the lab at 8 PM, before returning home at 2 or 3 AM. It appeared as though that routine would have to end in October with the start of the fall quarter, when I was scheduled to teach my morning undergraduate class. However, I found that Catherine would fall asleep in her snuggly as I walked to school and, if I went right into the classroom and began lecturing, she would stay asleep until class was over. So, I was able to care for my daughter considerably longer than I had expected.

With a crib set up in my office, after class I could check my mail and touch base with those in my lab before heading home with Catherine for lunch. Graduate students came to our apartment for research conferences, during Catherine's nap time, just as they had come to my office before she was born. Although the graduate students took my parenting in stride, none of my faculty colleagues visited during this period.

This arrangement was great while Catherine was an infant, but the single caretaker role became more and more difficult to sustain as she got older. We hired a sitter for a few days a week after Christmas, allowing me to teach the Field Ecology course and continue our field experiments. Although I published nothing the year Catherine was born, I had nine publications, more than double my average, the year after; a pattern which continued with the births of my two other children, Megan (b. 1990) and Travis (b. 1992). It proved to my skeptical colleagues that, despite enjoying parenthood and fully participating in childcare, I was still a "serious scholar."

Sociality in the Suburbs: Willow Leaf Beetles

The Field Ecology course was my antidote for the confines and tedium of the laboratory. Working daily in a cloud of frass and flour, I became very allergic to the beetles. Dave had developed similar allergies during his graduate research at Stony Brook. Once he had established a field research program, he used his remaining postdoctoral time in my lab to break his dependency on Benadryl inhalers by "going cold turkey — one nostril at a time." To control my own allergies, I alternated between sips of coffee and sips of children's Benadryl while counting beetles.

In the spring of 1979, in the Tinley Creek Forest Preserve, I discovered a population of willow leaf beetles, *Plagiodera versicolora* (Laicharting), a small metallic blue beetle whose larvae lived in groups (Figure 8.1 and Figure 8.2). They proved to be an ideal organism for studying the ecological genetics of primitive sociality and for calibrating the *Tribolium* laboratory studies of group selection against Nature.

Saplings of the sandbar willow, *Salix interior*, surrounded the vernal ponds of Tinley Park (Breden and Wade 1985; Wade and Breden 1986). I found it easy to bend and census entire saplings while the class conducted their own projects. Although the literature suggested that the black willow, *S. nigra*, was the beetle's preferred host plant, at our field site, the larvae fed exclusively on the leaves of *S. interior*, despite the availability of *S. nigra* and other willow species nearby.

Although the beetle has three or four generations per year in Illinois, the class field trips, beginning in late February, and my childcare schedule confined my early observations to the first generation, when the adult beetles emerged from hibernation, fed, mated, and began laying eggs. Ours

FIGURE 8.1 A laboratory clutch of willow leaf beetles hatching and cannibalizing some of their siblings in a large petri dish. The substrate is a wet paper towel and in the foreground is a piece of willow leaf.

was the first study of the natural history of *P. versicolora* (Breden and Wade 1985; Wade and Breden 1986) and our first publication combined data on over 8,600 individuals from four field seasons, 1979 through 1984. Felix and I used these early data as the foundation of an NSF proposal investigating kin selection rather than group selection.

The adult beetles overwintered under the bark and in the debris at the base of willows. They emerged from hibernation in late April to early May and fed on willow pollen and leaves, before mating and laying eggs. We rarely observed adults of this generation in flight and females laid eggs on leaves at the base of willows (Wade and Breden 1986; Wade 1994). Egg laying started most years between May 5th and 10th, except for the unusually cool spring of 1983. More than 98% of the time, females laid their eggs on the undersides of leaves of a single tree in clutches of two to 48 eggs, averaging 15 or 16 per clutch. In the field, females laid an average of two or three clutches, but in the lab, they would lay as many as 10. After overwintering, adult longevity in the spring was only 3–5 weeks.

Because the adults tended to aggregate on only a few trees, clutches were clustered. We often found more than half of all eggs on only 20% of available willows. Moreover, the timing of oviposition was synchronous, with females laying the large majority of eggs (84%–97%) within a week of our discovering the first clutch.

We used twist-ties and magic markers to indicate the base of the leaf where a clutch had been laid. We returned every day or so to count the eggs and the larvae emerging from them. Although the eggs developed quickly,

FIGURE 8.2 A willow sapling with eight clutches of eggs, each encased in a mesh bag made of bridal veil.

often hatching within four days, they were very vulnerable. Predators included warblers, lady beetles, and the larvae of hover flies (Syrphidae). Half of all clutches were completely destroyed by these predators and half of the remaining eggs were eaten by bugs or killed by parasitoid wasps; only 25% of eggs laid survived to hatch into larvae in Nature.

As if the array of predators were not risk enough, early-hatching larvae cannibalized other larvae and eggs of their own clutch. When we observed hatching in the field, we saw cannibalism 40% to 80% of the time, depending upon the year. Laboratory-reared clutches were no different (Figure 8.2). In the laboratory, with natural predators excluded, we found that as many as 45% of eggs and hatching larvae were lost to cannibalism. The period of cannibalism was brief, lasting about 24 hours, and after that time, larvae were strictly herbivorous, feeding on willow leaves.

After hatching and its attendant cannibalism, surviving larvae established a group feeding site on the undersurface of their natal leaf. Each larva locked its mandibles onto the leaf surface and rocked back and forth. When one broke through, the others aggregated around it and expanded the opening, lining up in a row or phalanx. Some of the smallest groups never successfully established a feeding site, and died trying.

Feeding rate depended upon group size and maturation required feeding on six or seven leaves. However, larvae did not move from one leaf to another as a group. Each moved alone, but left a chemical trail that subsequent larvae followed and reinforced. This resulted in the group reassembling itself on another leaf. Larvae did not move far, so it was easy for us to follow groups from one leaf to another, moving the twist-tie markers from old leaf to new leaf as they went. Group members molted at the same time and we would find their shed skins together on a leaf. When approached by insect predators, the larvae secreted salicylaldehyde, a noxious defensive substance, from eversible glands on their back. This behavior was an effective repellent against insect predators, and its effectiveness as a deterrent increased with group size.

Larvae foraged in groups for six days through two molts. After the third molt, group cohesion broke down as the large larvae entered a "wandering" phase. They dispersed throughout the willow canopy and molted to the pupal stage attached to the underside of a leaf. During the 4- to 5-day pupal stage, beetles were again vulnerable to predation, with only 34% developing successfully to adults. Some were parasitized by wasps, but larger numbers were eaten by other insects.

This life history made willow beetles the perfect field system for studying the evolution of social behavior. Here was an organism that lived in groups and interacted with its kin. When aggregated, the larvae displayed several primitively "social" traits, including synchronous molting, chemical defense against predators, and the ability, like ants, to follow the trails of other larvae. Like *Tribolium*, they were also at times intensely "antisocial" or cannibalistic. Thus, their kin structure had ecological and genetic consequences for the group as well as social and antisocial behaviors. We predicted that the ecology would influence selection within and between larval groups, while the genetics would determine the efficacy of each level of selection.

First and foremost, I wanted to measure the fitness advantage of living in groups and determine whether it was better to live in a large group than in a small one. And, if there was a fitness advantage to life in a large group, why did hatching larvae cannibalize potential group members, making their group smaller? To answer these questions, we needed to be able to manipulate group size and compare the success of groups of varying size and with different proportions of cannibals.

The Variable Advantage of Group Life

Natural Groups

Because we monitored hundreds of clutches from laying to hatching to dispersion each year, we could calculate how long a larva survived as a function of the initial size of its group. We found that, on average, larvae in large groups lived longer than those in small groups. Some small groups might be eaten shortly after hatching and disappear entirely, while others escaped discovery by predators and lived as long as larger groups. However, larvae in large groups rarely suffered the early losses to predation that afflicted the small groups. Unlike the constant group benefits in our mathematical models, the viability advantage of group living varied from year to year. In some years, *only* larvae in large groups survived. In other years, especially when predators were scarce, group size did not seem to influence larval viability: larvae in smaller groups survived just as long as larvae in larger groups. In theoretical models, the group fitness benefit was treated as a constant; in Nature, it varied from year to year. Not surprisingly, Nature was more complex than theory. Overall, it appeared that it was good for larvae to live as a group, especially early in life.

The correlation we had observed between group size and larval viability might not have been causal. The larval survival might be a simple reflection of their mother's health, if females in poor health laid fewer eggs, on poorer quality leaves, and the eggs then hatched into smaller, sicklier larvae. Conversely, healthy females might lay more eggs, on higher quality leaves, which then hatched into larger, more robust larvae. Maternal health, not group size per se, could have been the reason we observed an association between group size and larval survival. The only way to determine if group size *caused* larval survival was to conduct a manipulative experiment, in which we randomly assigned larvae to groups of different size. In this way, no matter how many eggs a female laid and no matter how sickly or healthy a larva at hatching, we would control group size and the social environment of early life.

Felix and I hypothesized that group living conferred two fitness advantages. First, it increased the efficiency of feeding, allowing larvae to develop faster. Secondly, it increased larval survival because a group was more effective at repelling predators than a single larva. We imagined that larger groups had a greater volume of defensive chemicals and therefore were better defended against predation. We could test these hypotheses by protecting groups of larvae from predation. When shielded from predators, the group-size advantage to survival should disappear while the group-size advantage for feeding efficiency would remain. We also thought we could use paper toweling to blot up secreted chemicals and thereby deplete group defenses.

To test our hypotheses or theoretical predictions, we had to conduct manipulations of larval groups in the field. Felix and I, with the help of the McCauley lab at Vanderbilt University, developed a willow-beetle study system that integrated laboratory rearing with field experimentation to address the ecology of selection and the genetics of the response to it.

Our primary field tool was a small mesh bag with a draw string bottom, which Carol Kelly sewed for us from bridal veil by the hundreds. By placing a bag over newly laid clutches we were able to exclude predators (Figure 8.2). Larvae could feed inside a bag without risk of predation. This technology also let us store clutches in the field and harvest eggs or larvae from them to use in our experiments. The bags also functioned as mating chambers. Placed inside a bag, a pair of virgin adults would mate readily, giving us control over matings, thereby allowing us to estimate heritability in the field.

Deer were a nuisance. For some reason, deer would chew on the mesh bags, carry them a short distance and spit them out. The larvae did not survive these encounters. We hung up bars of soap amid our bags as a deterrent. Although deer at first chewed on the soap too, they soon left our bags alone.

We also developed a laboratory system for rearing willow beetles. We used the laboratory colonies to develop genetic markers and resources for estimating the heritability of body size, clutch size, and cannibalism. Although willows grew like weeds in the forest preserves, we found them much more difficult to grow in the greenhouse. We had to import most of the "willow fodder" for our laboratory colonies from our field sites or from willows growing along the lakefront in Hyde Park. There were synthetic insect diets available for other types of leaf-eating beetles, but they proved insufficient for sustaining our laboratory colonies, even when blended into "shakes" with ground-up willow leaves.

Dave and his masters student, R. O'Donnell, developed electrophoretic assays for willow beetles, giving us genetic tools for estimating relatedness between larvae, including that between larval cannibals and their victims. Applying their protocols to our laboratory population, Felix developed a set of pure breeding lines, homozygous for one or another of several different, polymorphic allozymes. With these lines, we were able to create genetically distinguishable cannibals and noncannibals (see below) in the lab, transport them to the field, and establish larval groups with different known proportions of cannibals and noncannibals. Even better, we could determine whether the cannibals or the noncannibals enjoyed a survival advantage within groups by collecting the survivors and surveying their genetic markers.

Although the willow beetle system never approached *Tribolium* for ease of use and diversity of purpose, the methods described above let us manipulate enough of willow beetle ecology and breeding biology to measure selection within and between groups in the field. With this simple toolset, we were able to calibrate our theoretical and laboratory findings against Nature.

Group Size and Larval Growth Rate

We wanted to determine the effect of social context on larval viability and growth rate. Moreover, we wanted to separate the potential group effect on feeding and growth rate, from the group effect on viability via predator defense. Unfortunately, we were unable to measure individual growth rates in Nature, because we could not uniquely identify individual larvae and because they were so tiny. Only the sensitive electric balances in the lab were up to this task. However, using the "sacrifice" design of McCauley and Wade (1980), we could set up replicate groups in the field and then sacrifice them at daily intervals. We could take back those sacrificed to the lab for weighing (Breden and Wade 1987). By putting the weights of different replicates of the same group-size treatment in a time series, we could reconstruct larval growth trajectories in the field!

We set up our group size experiment like this. We found clutches in the field with at least 12 eggs and collected two newly hatched larvae (< 24 hours old) from each clutch for weighing back at the lab. These two larvae gave us size at hatching for each group and, in analyzing our results, they could be used as covariates to statistically remove the uncontrollable initial variations in body mass. The remaining larvae were randomly assigned to a group-size treatment, consisting of 2, 4, 6, 8, or 10 larvae. For example, with a clutch of 14 larvae, we would collect two and, if the group-size replicate was 8, we would collect an additional 4 larvae and leave a group of 8 larvae together in the field. In this way, all larval groups began life where they were laid but we controlled their group size. Because larvae from large clutches could be assigned to smaller group sizes but not vice versa, we were limited to investigating the lower end of the group size distribution (groups ≤ 10 larvae). Because these clutches were not protected from predation prior to hatching, we did not have an estimate of cannibalism rates in this experiment. In total, we set up 120 larval groups, 24 replicates of each group size. In order to concentrate on the effects of group size on feeding and growth rate, and eliminate ants and others predators, we applied a sticky substance, Tanglefoot, to the tree above and below each group. We conducted a daily census of all groups and sacrificially collected replicates of each. At the end of 6 days, we collected the remaining larvae as well as all the leaves they had eaten.

From our initial collections of two larvae from each group, we found that some larvae were larger than others at hatching and that these differences continued to affect feeding rates and growth rates throughout life. That is, larvae that were larger at the time of hatching fed more and grew faster than larvae that were smaller at hatching. We also found a conspicuous and independent group effect: regardless of size at hatching, larvae in the smallest groups did not feed as efficiently or gain weight as rapidly as larvae in the larger groups. Of further note: with some predators excluded by Tanglefoot, we found no differences in survival among the different sized groups.

Given the conspicuous effect of initial larval size on feeding and growth rates, we wanted to determine whether particular females laid large or small eggs and whether being cannibalistic conferred a size advantage to some larvae over others as it did in the lab (Breden and Wade 1985, 1987). But, to measure cannibalism in the field, we needed newly laid clutches of eggs protected from other sources of mortality (i.e., predation).

Group Size, Cannibalism, and Larval Viability

We collected mated females from the lab and field and placed them inside mesh bags on willow trees until they had laid a clutch of eggs. We then moved the female but left the eggs inside the bag where they had been laid. Since predators were excluded, we could attribute the loss of life at hatching to cannibalism and use this information in our subsequent analyses. For example, if the number of larvae at hatching was two less than the number of initial eggs, we could be certain that there was cannibalism in the group. However, we could not distinguish the case of one larva eating both eggs from each of several larvae eating some of one or both eggs. That is, our method could not really characterize individuals as being cannibalistic or not, even though, in our monitoring, we often witnessed cannibals in the act. In many groups, there was no cannibalism: all eggs hatched successfully into larvae. In many other groups, there was clear evidence of cannibalism, even though we could not identify the specific perpetrator(s). Once a group of larvae hatched, we removed the bag. We did not use Tanglefoot this time, and instead allowed natural predation to affect the groups.

Results

Figure 8.3 below shows our results. We separated our groups into two categories, those with and without cannibals. In the absence of cannibals, we found a clear advantage to living in groups (Figure 8.3 A): the larger the group, the greater the survival advantage. However, in groups with one or more cannibals there was no survival advantage to living in a group (Figure 8.3 B). Our field notes taken each day as we followed these groups throughout development told us why the effect of group membership differed for groups with and without cannibals. Cannibals had two effects that were detrimental to the other larvae in their group—besides eating some of them, an effect removed by our experimental protocol. First, cannibals disrupted the group synchrony at molting. The cannibals, by virtue of growing faster, molted earlier and headed off to find fresh leaves sooner than the noncanni-

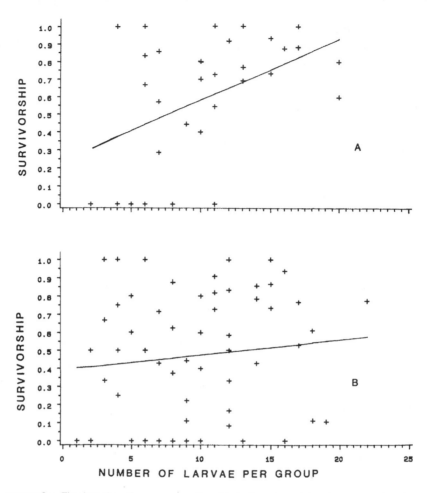

FIGURE 8.3 The data show the average survivorship to day 4 in each larval group as a function of the initial number of larvae in the group. There are two kinds of groups: (A) groups with no cannibals; and (B) groups with one or more cannibals. The lines are the linear regressions of survivorship on initial group size. Larger groups provided protection from predators for groups without cannibals, but this benefit was not enjoyed by the groups with cannibals. See text for further discussion.

bals in a group. The group's transition from an old leaf to a new one became less of an orderly single-file march, and more a chaotic sequence of independent transitions. Secondly, the early molting cannibals traveled farther distances from the natal leaf to the next fresh leaf, leaving the smaller, non-cannibalistic, group members spread out over a wider area. As a result, not only did they not arrive at the next leaf together; some were lost along the way, resulting in a smaller group. (Note that the x-axis in the Figure 8.3 is the initial group size and not the group size after some members had been lost.)

Lab data supported our field conclusions: cannibals enjoyed significantly accelerated growth rates relative to noncannibals. In other field experiments, we also discovered an interaction between group size, cannibalism, and survival. Larvae in the smallest groups of noncannibals did not survive, even in the absence of natural predation, because they fed so poorly and were sometimes unable to establish a feeding site. However, small groups of cannibals had no problems feeding (Breden and Wade 1987; Wade 1994).

Variability in Genetic Relatedness within and among Larval Groups

Relatedness within Larval Groups

Although we sometimes observed males and females mating in the field in spring, we did not know whether a given female mated once or many times before laying her eggs. We also did not know whether adult females mated before hibernation and stored sperm over the winter for use in laying in the spring. Multiple mating by females and subsequent multiple paternity of broods would reduce the genetic relatedness within a brood from the full sib value of ½ toward the maternal half-sib value of ¼. On the other hand, a female might add her eggs to the clutch of another female and thereby decrease relatedness even further, below ¼. With variable relatedness within broods, the cannibalism we observed in Nature might be directed by the cannibals toward the eggs and larvae to whom they were less genetically related, such as those of different fathers or of different females, as many kin selectionists predicted it should be (Hamilton 1964; Polis 1981, 1984; Mumme et al. 1983).

There were also possibilities in the other direction. If the parents were genetic relatives, r could be elevated by inbreeding, and the efficiency of group selection should increase as we had seen in our *Tribolium* experiments and in our mathematical models. We needed variable genetic markers in our field populations to investigate these possibilities and to estimate r.

McCauley and O'Donnell (1984) used the highly polymorphic gene phosphoglucomutase, or *pgm*, to survey the broods laid in the lab by field-captured females. Although their method could not detect all cases of multiple paternity, they found direct genetic evidence of multiple paternity in 40% of clutches and, using three different statistical genetic methods, estimated the *r* among larval groups and found it to lie in the range of 0.38 to 0.44. That is, larvae in the same clutch, on average, were somewhere between full and half siblings. In a later study, using the same genetic marker, Stevens and McCauley (1989) discovered that most females mated before they entered winter hibernation (diapause) and that most females used the sperm of more than one male to fertilize their eggs in the spring.

As we developed additional molecular markers, we extended our investigation of relatedness. In a tremendous, three-lab coordinated effort (Felix at the University of Missouri, Dave at Vanderbilt, me at Chicago), we characterized *r* for larval groups collected over three generations at three localities in Virginia and three localities in Illinois. By design, the localities within each state were only a few miles apart. Larval groups were collected at the start of the second generation in 1985 (indicated as 1985-2 in Table 8.1), and from the first and second generations in 1986 (indicated as 1986-1 and 1986-2, respectively, in Table 8.1).

Altogether, we had 15 collections and 15 separate estimates of *r*, which are arranged in ascending order in Figure 8.4 (after McCauley et al. 1988).

TABLE 8.1 The location (Virginia or Illinois, sites 1, 2 or 3), laying time of brood (year and brood number), number of larval groups, average group size, and tree species on which the groups of willow beetle larvae were found (McCauley et al. 1988).

Locality	Date	Groups	Group Size	Host Tree Species
VA-1	1985-2	17	7.5	*S. caroliniana, S. nigra*
	1986-1	14	7.4	
VA-2	1986-1	21	6.9	*S. alba*
	1986-2	24	9.5	
VA-3	1985-2	12	5.6	*S. babylonica, S. nigra*
	1986-1	17	8.5	
	1986-2	32	9.0	
IL-I	1985-2	34	9.9	*S. nigra*
	1986-1	44	6.7	
	1986-2	34	9.9	
IL-2	1985-1	39	8.2	*S. babylonica, S. interior, S. nigra, S. alba*
	1986-1	29	8.3	
	1986-2	37	9.4	
IL-3	1986-1	36	7.4	*S. interior*
	1986-2	34	9.6	

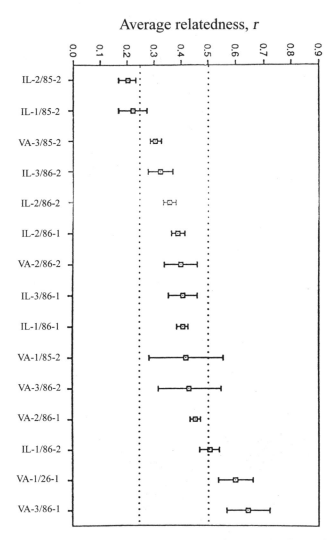

FIGURE 8.4 Depicted are fifteen, independent estimates of genetic relatedness, *r*, between larvae within groups from six localities and three generations. Each point is the value of *r* averaged over approximately 28 different several groups and the lines surrounding each box show the standard deviation of the estimate.

The dotted lines at 0.5 and 0.25 indicate the values of r for full- and half-sibs, respectively.

We discovered a great deal of spatial and temporal variation in r. Estimates of r spanned a three-fold range among years and localities. On the far left of Figure 8.4, the smallest r value was only 0.20, less than expected for maternal half-sibs. This meant that, on average, these larval groups consisted of offspring from more than one female. In contrast, on the far right, the largest two estimates exceeded 0.60, indicative of the unique signature of inbreeding: the parents of these groups were themselves genetic relatives. This variation was amazing to me. *There was more variation in r among natural populations of willow beetles than I had imposed in my kin selection experiment with Tribolium!* In fact, we were able to show that there were statistically significant effects on r owing to localities and to generation within locality, as well as a significant interaction between locality and generation. That is, r varied both spatially and temporally, because the underlying breeding biology of local populations also varied spatially and temporally. For willow beetles, r was not the unitary, species-specific parameter one found in kin selection and inclusive fitness theory and in discussions of the evolution of social behaviors. At the time, our repeated measurements of r for a single species were unique. Even today, such repeated estimates remain very rare (Ward 1983, ponerine ants; Costa and Ross 1993, eastern tent caterpillar; Estoup et al. 1994, honey bees) and I believe our study is still the largest study of spatial and temporal variation in r in a single species.

Variation among Larval Groups at Different Spatial Scales

We used these same data to investigate the hierarchical structure of F_{ST}, the population genetic differentiation. We had three levels in our hierarchy: between trees within localities, between localities within regions, and between regions. Locally, among trees, F_{ST} ranged from 0.01 to 0.10, very much in the range of values we had explored in our experimental metapopulations of *Tribolium* (Wade and McCauley 1984). This degree of very local population genetic structure exceeded that observed among localities within Illinois or within Virginia and was comparable to the regional difference, 0.057, between states.

If the fitness costs and benefits of group living were held constant (as they were in our models), this type of large, spatial variation in r could lead to the evolution of a social behavior in some places but prohibit its evolution in others, just as it had done in my kin selection experiment (Wade

1980a). This magnitude of variation in r could easily change the balance between group and individual selection (Wade 1980 a and d; Wade and Breden 1981; Breden and Wade 1981, 1985, 1987). Because of the temporal variation in r within localities, it would be possible for an antisocial behavior, like cannibalism, to be favored in one generation but not in the next. This sort of fluctuation in the predominance of the levels of selection might lead to polymorphism for cannibalism, with some individuals strongly inclined to eat kin while others were genetically disinclined to do so. Similarly, it was easy to predict that there should be geographic variation in the level of cannibalism, with lower levels in the regions with the higher values of r and higher levels in those regions where r was lower.

Our data and the beetle's ecology also raised the possibility of an interaction between selection and heredity in Nature, similar to the synergistic effects of our models. That is, it was possible to imagine that the same ecological factors that affected the fitness costs and benefits of group living also affected r. For example, a high local density of beetles might simultaneously enhance the efficiency of group defensive displays as well as increase the frequency of mergers between larval groups, lowering r. Alternatively, low local density might increase the risk of mortality for larvae and might also lead to inbreeding and increased r. This sort of temporal variation in the strength and direction of social evolution would certainly slow the rate of approach to a genetic equilibrium, if such an equilibrium even existed in Nature.

Multilevel Selection on Cannibalism

Group and Individual Selection on Cannibalism

Our laboratory studies showed us that cannibalistic individuals developed more quickly than noncannibals of the same age. Our field studies had shown that *average growth rate* was faster for survivors in groups with some cannibals, even though group cohesion was disrupted and viability was lowered. However, in these field studies, we did not know how many or which of the surviving individuals had been the cannibals at hatching.

To find out, we developed a method for creating cannibals and noncannibals in the lab. From a clutch of eggs, we would isolate single pairs of eggs. If a pair hatched into two larvae, then neither was a cannibal; if a pair of eggs hatched only into one larva, that survivor was a cannibal. (We used a microscope to make sure in the latter case that the nonhatching egg had been eaten as opposed to simply not hatching.) This procedure gave us an ample supply of cannibals and noncannibals to use in our field experiments. By obtaining cannibals from one homozygous line and noncannibals from another, we could set up groups in the field with different proportions of cannibals and noncannibals and distinguish them from one another using unique genetic markers. Although, in mixed groups, cannibals and noncannibals would mingle in the field, at a later time, we could collect survivors and use electrophoresis to sort them out again by their genetic markers. These methods allowed us to measure selection on cannibalism within and among groups in Nature, in a manner comparable to my earlier study of cannibalism with *Tribolium*. In addition, these studies gave us a valuable by-product: from the fate of our egg pairs, we were able to document the existence of heritable variation for the tendency toward cannibalism in willow beetles (Breden and Wade 1985).

We transported our cannibals and noncannibals of known genotypes to the field, where we established groups of 8 larvae with 25%, 50%, or 75% cannibals. We let these groups develop in the field for 6 days, collected the surviving larvae, and returned them to the lab. There, we weighed and electrophoresed them to determine genotype and initial status as a cannibal or a noncannibal.

In all but one group, we found that cannibals survived better than noncannibals and, in every group, they were heavier than the noncannibals. In fact, cannibals were more than twice as heavy six days after hatching as noncannibals. The small initial advantage in size of a cannibal increased 10-fold in a very short period of time, indicating that the early growth advantage of cannibalism was sustained and augmented by faster feeding and higher growth rates during later development.

We repeated this experiment varying *both* group size and proportion of cannibals, by establishing groups of 2, 4, or 8 larvae with 0%, 50%, or 100% cannibals. Here, we found that survivorship was higher in groups without cannibals than in groups with cannibals, similar to our earlier study with "natural" groups. And, if all members of a group were cannibals, group synchrony was no longer disrupted as we had seen earlier in mixed groups of cannibals and noncannibals.

Group Selection in Laboratory, Theory, and Field

Taken together, our willow leaf beetle studies indicated that individuals benefited from living in a group in terms of survival and development.

However, the presence of cannibals within a group diminished or eradicated these group advantages. Why did larvae cannibalize relatives, if membership in a large group enhanced survival? Our data suggested that it was the evolutionary result of conflict between two opposing levels of selection, a conflict that was irreconcilable even in principle owing to continual spatial and temporal variation in group benefits, individual costs, and genetic relatedness, the constants of kin selection theory.

Individual selection favored cannibals because it enhanced early larval growth. Failure to establish a feeding site was one of the most important causes of mortality of first instar larvae in the field. Cannibals grew more rapidly and their larger size conferred upon them a mechanical advantage over a tough leaf surface; they did not need a group to initiate or to continue feeding. In contrast, group selection opposed cannibalism. Cannibals disrupted the synchrony and cohesion of groups. This reduced the group size and diminished its survival advantage in the face of predation and its feeding rate advantages. Like the individual fitness effects of cannibalism, the fitness effects of living in a group in Nature were sometimes very large. For example, in one study, the proportion of larvae surviving to day 4 in groups of size 2 was only 3%, while it was nearly twenty times greater, 50%, in groups of size 10. Thus, the two levels of selection tended to act on cannibalism in opposing directions.

In the field, there were two important sources of ecological variation acting to sustain the conflict between the two levels of selection. First, the individual fitness benefits and the group fitness costs to cannibalism varied spatially and temporally, unlike the constant fitness effects assumed in the mathematical models (ours as well as those of others). In some generations in some years in some localities, there was no apparent advantage to life in a group. Under those circumstances, cannibalism was favored by individual selection and unopposed by group selection. At other times in other places, predation was strong and group membership was essential to survival; cannibalism disrupted group cohesion and was opposed by group selection. Indeed, even the host species of willow where mom beetle laid her eggs might play a significant role in the effectiveness of larval defensive displays in repelling predators. On the long thin leaves of *S. interior* and *S. babylonica*, larval groups were arranged in lines, with larvae at both ends more vulnerable to predation than those in the middle (Wade 1994). On the more rounded leaves of *S. nigra* and *S. alba*, larvae fed in more circular groupings (called *cycloalexy*, Jolivet 2008), presenting a more uniformly effective defensive display, protective of all group members.

Second, whatever the strength of selection within or among groups, the heritabilities, h^2 and g^2, changed spatially and temporally because r did. In some places in some generations, inbreeding was high, so that selection among groups was particularly effective relative to selection within groups. In other generations in other places, females mated multiply and added eggs to the clutches of others, lowering r. As a result, selection among groups was particularly ineffective relative to selection within groups. Willow beetles in Nature expressed the same type of variation in individual and group heritability that we had imposed in our laboratory experiments with *Tribolium* (Wade 1980a) and explored with our theory (Wade and Breden 1981; Breden and Wade 1981).

In Summary

The results from our in vitro experiments with *Tribolium* were supported in every respect by our in silico studies using mathematical models and computer simulations. The in vivo field studies with the willow leaf beetles went further. They revealed unexpectedly high levels of variation in both fitness and heritability at each of the two levels of selection. Nature was much more complex and more extreme in her treatments than we were in the laboratory. Natural populations were much more variable than theory imagined.

The results of our flour beetle experiments with cannibalism and genetic relatedness were abundantly supported by results from our decade-long willow leaf beetle studies. Willow leaf beetles could be just as cannibalistic as *Tribolium* and both species exhibited considerable heritable variation in the tendency toward cannibalism. As far as "calibration" was concerned, the *Tribolium* laboratory system fit comfortably within the range of population structures, individual and group fitness effects, and genetic behaviors displayed in Nature by willow leaf beetles.

Even the hazards of the field were comparable to those of the lab. In the lab, we endured the allergies and the mind-numbing boredom of counting beetles. In the field, we worked knee-high in sewage and fought clouds of mosquitos with heavy flannel shirts and blankets, while we collected, counted, and marked larvae. During this period, I felt that my lab had achieved the Triple Crown of population biology: good lab work, good fieldwork, and good theory.

9

Experimental Studies of Wright's Shifting Balance Theory

Early Discussions of Interdemic Selection with S. Wright

At the end of my first year as an assistant professor at Chicago, Park made arrangements for me to meet with Sewall Wright at his home in Madison, Wisconsin, on May 20, 1976. Wright sent me a letter with a hand-drawn map of the walking route to his home from the University of Wisconsin campus. Park encouraged me to send Wright an advance draft of the main paper of my thesis, so that I could benefit from his guidance on future experimental work. Wright was a gracious host, boiling us chicken noodle soup for lunch. In our discussion, Wright emphasized that my experiment differed from his theory in the mechanism of group selection, which he preferred to call interpopulation or interdemic selection instead of group selection. Wright's mechanism of interdemic selection was differential dispersion, wherein populations of high fitness sent more migrants out into populations of lower fitness and populations of low fitness sent out few or no migrants at all. In contrast, I had imposed group selection by the differential extinction of populations. Wright found my results interesting but not surprising. He explained that he considered differential extinction with colonization a more effective form of interdemic selection than the differential dispersion of his Shifting Balance Theory (SBT).

In our discussion, I suggested an experimental design that would more closely resemble Wright's differential dis-person process, a group selection experiment with migration proportional to deme productivity instead of extinction. Wright thought that I would have to have at least 10,000 experimental populations in order for random drift to find a favorable gene combination for fitness. Although he believed that gene interactions were ubiquitous, he also believed that deleterious gene combinations were more common than beneficial ones. He considered good gene combinations to be so rare that discovering one by drift would require extensive sampling and, hence, a very large number of small populations. Any experimental study of his process would also have to allow sufficient time (perhaps considerable time) for chance to assemble a favorable gene combination. These requirements would necessitate an experiment two orders of magnitude larger than all the experiments of my thesis put together. Moreover, the experiment would need to be run for several years or longer. Wright and I discussed only epistasis as a cause of population genetic differentiation. We did not discuss the possibility of genes with indirect effects of the sort that McCauley and I discovered a few years later.

Wright also expressed his concern that the difference in fitness between a population with a favorable gene combination and one without it would probably be small and thus hard for me to detect statistically. That is, on average, a "good gene combination" would be better, but not great, for mean population fitness. As a result, even after a favorable gene combination arose by chance, only very modest levels of interdemic selection by differential dispersion would ensue. Wright cautioned me that the early response to interdemic selection might be so long in coming and so weak in magnitude as to escape my detection. An inadequate experiment with weak results would not constitute a real test of his theory; nor would the absence of a result refute it. All in all, I found ours a very discouraging discussion in regard to my ambitions for using the *Tribolium* system to test Wright's grand theory.

Although I believed I could do anything with flour beetles, as a practical matter, 10,000 populations was 20 times more than the maximum number Ora Lee and I could handle. And, I would need at least twice that number of populations to include the necessary control metapopulation with random migration instead of differential migration. Replication of both the experimental and control metapopulations only magnified the impracticality. The cost too would be prohibitive because, even with 800 populations per incubator, I would need dozens more incubators than I had requested in the NIH proposal then under review. Moreover, there was no way that that many incubators would fit into the lab space I had just been assigned by the department chair.

For all of these reasons, I shelved the idea of experimentally testing Wright's Shifting Balance Theory (SBT) for more than a decade. It was not until after tenure that I thought I could take the risk of setting up a very large experiment that might ultimately fail. By that time, I had also been emboldened by the steady stream of positive group-selection results from our much smaller *Tribolium* metapopulation experiments and from the field studies with willow leaf beetles. Both laboratory and field studies indicated that heritable variation among demes was not going to be a problem and that, given sufficient drift, among-deme fitness differences could be expected to appear rapidly and be readily detectable. Furthermore, our results pointed to genes with indirect effects affecting traits like cannibalism, instead of Wright's epistatic gene combinations, as the causal basis of the response to group selection. This eliminated, in part, the problem of a lengthy waiting time for drift to assemble a favorable gene combination. We had seen drift create g^2 rather rapidly in all of our experiments. In none of our earlier work did we have to wait more than three or four generations to observe substantial heritable variation in mean fitness among our demes. There was much more genetic variation for mean fitness latent within our experimental populations than Wright (or anyone but Park) had expected.

In 1987, an opportunity arose that allowed the anticipated work load of a test of Wright's Shifting Balance Theory (SBT) to be divided: I found an enthusiastic partner for counting beetles in my former student Dr. Charles Goodnight. At this time, Charles was a postdoctoral fellow with Dr. David Mertz at the University of Illinois at Chicago, where he was conducting other flour beetle experiments, including his now well-known community selection experiments, the first of their kind (Goodnight 1990 a and b). However, during this period, he frequently came to my lab to visit his family. His partner, Dr. Lori Stevens, was my postdoctoral fellow and we had set up another crib in my nonbeetle space for their infant daughter, Lisa, just as I had done earlier for my daughter Catherine. Since a test of the SBT was an "extra" experiment, added on to the other work that we each were already doing, Charles and I both needed to commit to it to make it a feasible undertaking. Together, we could do a much larger experiment than either one of us could have done alone. Per Wright's earlier admonition, we both believed that "size mattered" in this experiment. Although a decade earlier I had spoken about testing the SBT with Wright, the specific protocol for our experiment was the result of a long brainstorming effort between Charles and myself.

In the following sections of this chapter, I will first sketch Wright's Shifting Balance Theory (Wright 1931, 1932) and the problems we had to solve in order to experimentally test its predictions. Then, I will describe the design that Charles and I devised to test its efficacy. The findings from the first three years of that study were published in 1991 (Wade and Goodnight 1991) but I continued the experiment, at a somewhat reduced scale, without Charles for another four years. I will present some of those additional data here. In chapter 10, I report the results of our investigations into the genetic basis of the response to interdemic selection and the replications of the SBT experiment, conducted after Charles departed for a faculty position at the University of Vermont, where he is currently a full professor.

Gene Interactions

Wright's Shifting Balance Theory (Wright 1931, 1932) was founded on several empirical observations and the deductions about the evolutionary process that Wright drew from them. In complex interacting systems, Wright argued that some gene combinations would have fitnesses equal to one another while, at the same time, other gene combinations would have higher or lower fitness, much like some hands having better or poorer card combinations in poker. In a metapopulation, gene combinations good for fitness would become fixed by drift or selection within some demes. Even a relatively poor gene combination might be the best combination available for individual fitness locally. This meant that, across a metapopulation, local demes would manifest a multiplicity of "adaptive fitness peaks," which Wright termed *local fitness optima* (Figure 9.1 below reprints his famous figure [Wright 1932, p. 358]). With many interacting genes, the number of possible peaks increases geometrically with the number of genes (Wright 1969). Without interactions between genes or between genes and environments, there would be only a single peak and no need for a process allowing for the evolution of demes from one fitness peak to a higher one.

Individual selection, which Wright called *mass selection*, created the local fitness optima by moving each deme toward the nearest fitness peak in genotypic space and holding it there. If population size were sufficiently small relative to the strength of local individual selection (Wright 1931), the deme might move across a fitness valley separating its current peak from the next nearest peak by random drift. However, the *nearest* peak was un-

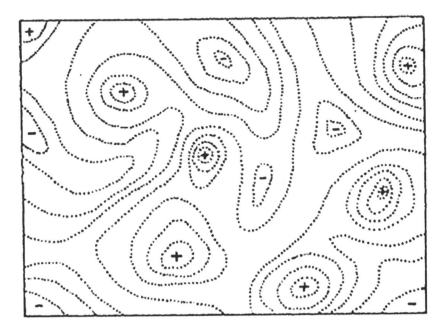

FIGURE 9.1 Wright's depiction of an adaptive landscape. The fitness peaks are identified by +
and the fitness valleys between peaks by –. The fitness contours connect points of equal mean
fitness (Wright 1932).

likely to be a higher fitness peak, let alone *the highest* peak, given the myriad
of possible optima afforded by gene interactions. How, then, could a species
find the best among good? Wright added interdemic selection to his SBT to
answer this question.

Just as Darwin used artificial selection by breeders as a model for his
theory of evolution by natural selection, Wright used the genetic transformation of domesticated longhorn cattle by breeders as a model for his SBT.
USDA breed histories showed Wright that, although local breeders strove
to breed the best animals, local selective efforts were not the primary force
transforming a breed. The transformation occurred when, by a combination
of chance and local selection, a local farm produced an important combination of genes. Subsequently, demand by other breeders on other farms for
that improved quality of animal led to the dispersion of the favorable gene
combination out from the originating farm and into other farms. Interdemic
selection in Wright's SBT was Nature's version of this process of breed transformation. Interdemic selection by means of differential migration occurred
as good gene combinations were exported out from demes of high mean fitness and into demes held at lower fitness peaks by local individual selection.

Why should populations with higher mean fitness send out more migrants than others? Wright proposed an ecological mechanism wherein individuals moved out from crowded areas of high density, where competition was intense, and into more sparsely populated areas of lower density with less competition for resources. His idea was founded on a theoretical relationship between mean absolute fitness, W, and changing population size, N—namely, $N_{t+1} = WN_t$. That is, the size of a population at one generation equals its size at the previous generation multiplied by mean fitness. According to Wright's chain of logic, good gene combinations caused high mean fitness; high mean fitness caused high population density; and the competition that attended high density caused emigration. Wright's argument connecting genes to ecology resonated with Park in their discussions.

Wright's metaphor of an adaptive landscape with a multiplicity of local optima is not just a genetic concept; any complex, highly interactive system of equations tends to have multiple solutions. Software researchers have learned how to "... 'breed' programs that solve problems even when no person can fully understand their structure" (Holland 1992). A famous practical, complex problem is the "postman problem," also known as the "traveling salesman problem": given 40 cities, what is the shortest possible route for visiting each one? Starting in one city and then going to the next nearest city is one solution, but it is not the best solution. In fact, there are so many possible solutions it is difficult to determine *if* there is a best solution, let alone "*the*" best solution. The current method for solving such complex problems is to use computer simulations and versions of Wright's theory called genetic algorithms (cf. Holland 1992 or "Sewall Wright meets artificial life" by Toquenaga and Wade 1996). With many computers working in parallel on the same problem and sharing "local" solutions among computers, it is possible to arrive at a best answer. Another example of a process that mirrors Wright's SBT is social learning, wherein students work independently on a complex problem and then share, compare, and contrast their solutions. Social learning is considered by some a fundamental process essential to human learning (Rittle-Johnson and Star 2007).

Gene Interactions Create the Fitness Landscape

Wright (1931, 1932) introduced his concept of the "adaptive landscape" or "surface of selective value" (see Figure 9.1) with the goal of elucidating how the simultaneous action of several evolutionary forces (mutation, migration, random genetic drift, and natural selection) resulted in adaptation and speciation. In Wright's words (1931, p. 101), "Selection, whether in mortality, mating or fecundity, applies to the organism as a whole and thus to the effects of the entire gene system rather than to single genes. A gene which is more favorable than its allelomorph in one combination may be less favorable in another. Even in the case of cumulative effects, there is generally an optimum grade of development of the character and a given plus gene will be favorably selected in combinations below the optimum but selected against in combinations above the optimum. Again the greater the number of unfixed genes in a population, the smaller must be the average effectiveness of selection for each one of them. The more intense the selection in one respect, the less effective it can be in others. The selection coefficient for a gene is thus in general a function of the entire system of gene frequencies." Wright argued that gene effects change as the genetic background changes, so the selection coefficient is not a property of the gene itself, but rather an abstract feature of the highly interactive genetic system. For Wright, genetics was more like the game of poker than like the game of war. It was Wright's focus and emphasis on gene interactions that shaped his SBT.

In later writings, Wright (1969, pp. 419–420; my emphasis in bold) elaborated on the "inadequacy of the simple additive concept of gene effect," stating that "all genes that approach additivity in their effects on varying characters will be favorable in some combinations and unfavorable in others in terms of natural selective value (fitness) and, thus, exhibit interaction effects of **the most extreme sort** in the latter respect." The focal trait of Wright's theory is *fitness* and the fitness of a genotype is not the sum of the properties of its component genes but rather an indivisible property of the entire genetic system (sensu interactions in statistics, which cannot be assigned as main effects to distinct but interacting factors). This foundational premise of Wright's SBT is different from standard theory, where Fisher's average gene effects came to be treated as constants, assigned to new genes when they arise by mutation (Williams 1966; Loewe and Hill 2010). Here, variations about Fisher's average are ignored (Williams 1966; Crow 2010; Hill et al. 2008).

The rate of gene evolution changes when a gene's effect changes in magnitude, and the direction of evolution changes when its effect on fitness changes sign (Wade 2001, 2002). In his SBT, Wright argued that interaction effects that changed the sign of a gene's effect on fitness were common (Wade 1992). In small populations, the existence of such interactions limits the domain of standard theory to those circumstances where the genetic background and environment remain constant. Thus, the evolutionary implications of gene interactions are two-fold. Interactions give the fitness landscape its irregular, bumpy shape, and they cause local variability in the rate and direction of adaptive change. Interactions create the possibility of group selection while simultaneously limiting the spatial coherence of individual selection.

The chapter 1 metaphor of the card games, poker and war, is illustrative. In war, there are no interactions; it has a fitness landscape with a single peak, and the highest card always wins. Interactions are the essence of poker; it has a rugged fitness landscape with many possible winning hands depending upon the combinations of cards, the spatial and temporal order in which hands are played, the abilities of other players in the game, and the past history of hands played by a "population" of players. In my metaphor, Wright's SBT is better than Fisher's theory as a description of Nature because there, too, interactions predominate, creating biodiversity at a local scale.

Wright's Shifting Balance Theory

Wright's SBT (Wright 1931, 1969, 1977; Wade and Goodnight 1991; Wade 1992, 1996; Johnson 2008) is considered to have three phases. Phase I is the "exploratory phase," where different gene combinations arise in different local populations as a result of random drift (chance). Wright believed that Phase I was necessary for discovering new adaptive gene combinations. Phase I is analogous to the shuffling and dealing phase of poker, which creates new card combinations. In Phase II, favorable gene combinations spread to fixation locally by the action of natural selection within populations. Because local populations are attracted to different fitness peaks, they become genetically more different from one another than they would by drift alone. The genetic differentiation can be especially large when interactions between genes change the sign of a gene's fitness effect, so that it is favored by selection in one population but eliminated by selection from another. Phase II is analogous to the "draw" stage in poker, where each player has an opportunity to discard some cards and replace them with others in an attempt to selectively improve the hand they were dealt by chance. Like a gene's fitness effect in different backgrounds, a card discarded by one player might be highly desired by another. Phase II results in a multiplicity of fitness peaks separated by valleys of lower mean fitness as each local population arrives at a fitness optimum and is held there by natural selection.

Wright (1932, pp. 358–359) considered evolution to this point incomplete: "The problem of evolution as I see it is that of a mechanism by which the species may continually find its way from lower to higher peaks. . . . In order that this may occur, there must be some trial and error mechanism on a grand scale by which the species may explore the regions surrounding the small portion of the field which it occupies. To evolve, the species must not be under strict control of natural selection." Phases I and II permit a species to hold on to selected gains while exploring neighboring genotypic space by random genetic drift for even better gene combinations. In a metapopulation, Phases I and II are a "trial and error mechanism" but together they do not guarantee a transition toward higher fitness peaks. In the metaphor of a complex problem with many solutions, each population becomes a small experiment in adaptive evolution (Wade and Goodnight 1998). Each population finds a local solution by the combined influence of chance and selection. The transition to higher peaks in Wright's theory requires an additional mechanism, such as that transforming the breed of shorthorn cattle.

In Phase III, migrants disperse out of populations with higher mean fitness and into populations of lower mean fitness. This is group selection by the mechanism of differential dispersal instead of differential extinction as in Wade (1977). Just as the fittest individuals send the most offspring to the next generation, the fittest populations send the most migrants into other populations. And, the migrants bring favorable gene combinations with them. Once a favorable gene combination arrives in a population at a lower fitness peak, selection, recombination, and recurrent migration together effect its transformation. By analogy with semi-conservative recombination, Phases I and II permit populations to hold onto selected gains, while still casting about in neighboring space for other fitness peaks; Phase III biases this exploration process toward the higher peaks. This was Wright's solution to the problem of adaptive evolution in the face of ubiquitous gene interactions. It was a solution that he found in the USDA breeding records of shorthorn cattle.

A Competing Theoretical Framework

Although Wright's SBT was influential (e.g., Dobzhansky 1937, p.190; Mayr 1942, p. 263; Simpson 1944), R. A. Fisher (1918) offered a competing theory that treated random genetic drift and gene interactions differently. Fisher (1918) used the average effect of a gene on fitness in his model of evolution by natural selection. Fisher's average (1958, pp. 30–31) encompassed all interactions, genetic and environmental. He believed that a population in Nature "comprises, not merely the whole of a species in any one generation attaining maturity, but is conceived to contain all the genetic combinations possible, with frequencies appropriate to their actual probabilities of occurrence and survival, whatever these may be, and if the average is based upon the statures attained by these genotypes in all possible environmental circumstances, with frequencies appropriate to the actual probabilities of encountering these circumstances." This average over all genotypic and environmental combinations, realized and conceivable, depicts adaptation by natural selection as a genetic process with little or no heterogeneity.

Astronomically large populations are necessary to achieve "all genetic combinations possible" and to have each genotypic combination occur in its global frequency in "all possible environmental circumstances." Dr. W. Ewens (2000, p. 33), a Fisher scholar, has remarked that "Fisher never paid much attention to the concept [effective population size] as he should have . . . and used extremely high population sizes (up to 10^{12}) in his analyses, surely far too large in general." For perspective, the value cited here is 100 million times larger than the best estimate of N_e for our own or any hominin species during its evolutionary history. However, even 10^{12} may not be sufficient for Fisher's average, given our current understanding of allele numbers per gene, of gene numbers per genome, and the number of microenvironments.

Wright placed gene interactions at the center of his theory, while Fisher placed them at the periphery of his. In order for interactions to be as important as Wright asserted, local populations must become different from one another in genetic or environmental background. For interactions to be as unimportant as Fisher asserted, a species must behave as though it were one large population with a homogeneous genetic background. Wright's theory relies on genetic drift to create local differences and he imagines that the members of a species are distributed into many, semi-isolated local populations, each with small N. Just as each farmer conducts a small experiment on his/her local farm, and each poker player conducts a small experiment on his/her dealt hand, natural selection operates locally within each small population. And, just as the efforts of different farmers result in the different outcomes, owing to variations in genes, feed, and practices among farms, natural selection interacts with the local environment, resulting in different genetic outcomes among local populations. In contrast, Fisher assumes a very large N so that differences in genetic background from one population to another are vanishingly small. One population is like any other with respect to natural selection and, on average, interactions exist but are irrelevant. Fisher's approach remains the foundation of today's standard evolutionary theory (Crow 2010; Hill et al. 2008), although, in small local demes, the variation about his grand average becomes too important to neglect, as our experimental work has abundantly demonstrated.

Dr. James Crow (1991, p. 973), an evolutionary theorist and longtime colleague of Wright, expressed the difference between Wright and Fisher this way: "Wright never showed much interest in experimental tests of his theory; his arguments were based on plausibility and analogy. He thought that much of evolution, the steady improvement of adaptation, could happen by mass selection acting on the additive component of the genetic variance, as Fisher said. But he thought that evolutionary creativity demanded something more. This might not happen often, and hence would be difficult to test in nature, but would be important when it did happen. . . . Fisher was interested in the steady improvement of fitness; Wright, in the occasional incorporation of novel gene complexes." Many concluded that Wright's theory could not be tested, because, like studying a lottery, it is difficult to characterize the rare or occasional events that distinguish his theory from Fisher's.

In response to my review (Wade 1980e) of the fourth volume of his magnum opus (Wright 1978), Wright wrote (Figure 9.2) that I had presented his SBT "with great clarity." He went on to express his frustration with the many misinterpretations of his theory "throughout the 50 years since I first presented it" (Figure 9.2).

The *Tribolium* Model of Wright's SBT

In 1987, although many features of Wright's SBT were supported by evidence from natural populations, almost 60 years had passed and it had not yet been subjected to direct experimental testing. Park, Goodnight, and I thought the *Tribolium* system was an ideal system for such a test. I was convinced by our earlier results that group selection capitalized on components of genetic variation not available to individual selection. And, I was no longer worried that the effects of one episode of group selection could be undone by one or even several episodes of individual selection without group selection. The cumulative evidence from our nearly 50 different and independent group-selection experiments had shown that the process worked across a variety of metapopulations and across a wide range of genetic subdivision ($0.03 < F_{ST} < 0.25$). Once the among-population genetic variation was in place and g^2 was high, our results showed that a single episode of group selection could

University of Wisconsin–Madison

LABORATORY OF GENETICS
Genetics Building
445 Henry Mall
Madison, Wisconsin 53706   January 18, 1980

Dr. Michael J. Wade
Department of Biology
University of Chicago
1103 East 57th Street
Chicago, ILL 60637

Dear Dr. Wade:

    I appreciate very much your review of my Volume 4. I have in the main had favorable reviews of this and earlier volumes, but with very few exceptions ·they were favorable for reasons that had little to do with my central thesis, which you present with great clarity.

    I especially appreciate this because my theory has unusually been wholly misrepresented throughout the 50 years since I first presented it. Huxley, in his "Evolution, the Modern Synthesis," 1942, thought that I considered random drift (which he, very embarrassingly to me, called "The Sewall Wright Effect), an alternative to natural selection, responsible for apparently nonadaptive differences between related species, an idea which he liked (though I had repudiated it). Fisher and Ford in 1947 made the same misinterpretations but strongly disliked it. Mayr, 1959, thought that the theories of Haldane, Fisher and myself of about 1930 were all the same, "bean bag genetics," fixation of rare major mutations one at a time. This applies to some but not all of Haldane's papers, not at all to Fisher's or my theories. He states that the Mendelian interpretation of quatitative variability was due to Mather about 1941. He evidently thought that because of this, any earlier theory was necessarily of the "bean bag" sort and that it was not necessary to read it before criticizing.

    Most recent authors have either taken the Huxley-Fisher-Ford view, probably at second hand, including a text book published last year by Futuyma or have ignored my theory as in the cases of Maynard Smith, Williams and Dawkins. Your quotation from Williams brings out very well the difference in viewpoint.

                            With best regards,

                            Sewall Wright
                            Emeritus Professor of Genetics

SW:gjw

FIGURE 9.2 Letter from Wright to Wade on January 18, 1980.

produce large and striking results. And, those responses endured, despite opposing individual selection, for years after group selection was relaxed (Wade 1984a). Although Wright's Phase III migration was group selection by the mechanism of differential dispersion rather than extinction, Charles and I were convinced it too would produce measurable effects.

The SBT Experimental Design

Unfortunately, we could not rely on Wright himself for a prescription for testing his theory. The most general prediction made by Wright was that the rate of evolution for traits like fitness with strong gene interaction effects would be faster in a metapopulation than in a large, randomly mating population. Wright's prediction was a qualitative one at best. He was not specific about the amount of gene interaction or the degree of genetic subdivision necessary to support accelerated adaptation. Nor did he specify the optimal values of N and m, or the strength of Phase III group selection relative to individual selection (see also, Wade 2013) necessary to achieve a significant acceleration in adaptation.

Not surprisingly, Wright's prediction of faster evolution in a genetically subdivided population had already received a great deal of experimental attention from animal breeders. However, their experiments had imposed artificial group selection on the wrong kind of traits. Reviews of artificial interdemic selection experiments (López-Fanjul 1989; Hill and Caballero 1992) found that Wright's process did not enhance the response of additive traits to selection. Because the traits under selection did not involve epistasis (gene interactions), these experiments were not considered an adequate test of Wright's SBT. As López-Fanjul (1989) concluded, "it is desirable that experiments of this kind [metapopulation selection] be carried out for epistatic traits in order to test Wright's hypothesis." In addition, the experimental protocols of these studies had pooled individuals across groups at every generation, thereby destroying g^2. Thus, these experiments gave us little insight or guidance for designing our own experimental test of Wright's SBT.

We knew that fitness was going to be our group-selected trait, but we had to decide how many demes there should be in our metapopulations, the number of breeding adults there should be per deme (N), the rate of migration among demes (m), the strength of the Phase III interdemic selection, and the nature of the control and amount of replication.

We created metapopulations consisting of 50 populations, each founded with 20 adults. Thus, each metapopulation would be a total of 1,000 breeding adults, subdivided into local groups connected by migration. Standard theory predicted that, if connected by any migration at all, such a metapopulation should evolve as though it were a single large, randomly mating population.

We converted Wright's verbal description of Phase III interdemic selection into an experimental protocol in this way. At the end of a 60-day generation, we had a census count, W, for the number of adult offspring produced by the 20 founding adults of each deme in each metapopulation. We did not impose differences in W among demes; the beetles created those themselves when one group of 20 adults was more or less productive than another. The beetle-created variation in W, however, did determine the level of Phase III migration in the experimental metapopulations and the level of random migration in the corresponding control metapopulations. We converted the census number into a *relative population fitness* value, W/W_{mean}, by dividing each W by the average (W_{mean}) of all fifty W values in the metapopulation. As a result, a highly productive population, where $W > W_{mean}$, had a relative population fitness value greater than 1 because it was more productive than average; and a population producing fewer offspring than average, where $W < W_{mean}$, had a relative fitness less than 1. We multiplied each relative fitness value by 20, the number of founding adults (N). For the more productive populations, $20 \times (W/W_{mean})$ was a number greater than 20; for the least productive populations, it was a number less than 20. Because of our definition of relative population fitness, the average value of the ratio (W/W_{mean}) equaled 1. For this reason, the sum of the 50 values {20 × (W/W_{mean})} was exactly 1,000, the number of adults we needed to set up the next generation.

We used the $20 \times (W/W_{mean})$ values to determine migrant numbers in this way. Consider two populations, one whose product ($20 \times [W/W_{mean}]$) is 23 and another whose product is 19 (see Figure 9.3). For the 23 population, we took 20 adults from it at random and used them to reestablish that population. We then took *another* 3 adults from it and placed them into a pool of migrants. For the 19 population, we would take out 19 adults at random and *added* to them 1 more adult from the migrant pool to bring the total number of founders to 20 adults. Each population with a relative fitness greater than 1 (or, equivalently, a product greater than 20) *contributed* migrants to the migrant pool; and, the most productive populations contributed the most migrants. Each population with a relative fitness less than 1 (or, equivalently, a product less than 20) *received* migrants from our migrant pool. And, because of the way we had defined population relative fitness, the total number of migrants received exactly matched the total sent.

Per Wright's verbal description of Phase III, the most productive populations sent migrants out into the least productive populations. Furthermore, our definition of *population relative fitness* paralleled exactly the standard definition and meaning of *individual relative fitness* used in population genetic theory. With this protocol, the variance in population relative fitness, V_w, equals the variance of productivity (variance of the W values) divided by the square of the average W. This ratio is called *the opportunity for group selection* (Crow 1958, 1962; Wade 1995). It sets an upper limit on the possible evo-

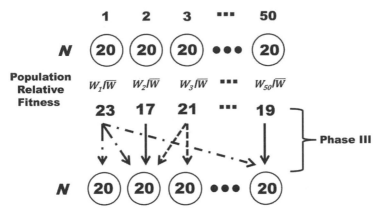

FIGURE 9.3 A schematic diagram of our protocol for an experimental metapopulation in our test of Wright's SBT.

lutionary response to Phase III group selection because not all of the variance among populations in W is genetic (Wade and Griesemer 1998); some of it is environmental. In this way, we could draw a direct connection between our experimental manipulations and Wright's evolutionary theory as well as a connection between concepts, like relative individual fitness, from standard theory and population fitness from Wright's theory. Importantly, if there were variation in W and if it had a genetic basis $(g^2 > 0)$, our experiment might work. If either was missing, it could not.

A Control for Phase III Group Selection

We needed a "control" metapopulation, where everything was the same as in our experimental metapopulations except for Phase III migration. Importantly, Charles and I wanted the amount of *random* migration among control populations to equal the amount of *Phase III* migration in our experimentals. We wanted the *only* difference between them to be in the *pattern of migration*, not in the number of migrants. Specifically, in our control metapopulation, migration would be random with respect to productivity, W, while in the experimental metapopulation migration was differential by W.

A control was not just good experimental practice; it was *essential* because a major argument against the SBT was that Wright's Phase III was unnecessary. Some argued that, with random migration, a good gene combination could get into any deme in a metapopulation and, once there, natural selection would insure its spread (formalized in Barton 1992; Barton and Rouhani 1993). If random migration and individual selection could spread good gene combinations, then Wright's Phase III migration would not add much, if anything, to adaptation; it was superfluous. Our experiment would put this argument against the SBT to the test. The standard argument would be supported if mean population fitness, W, increased over time in *both* our control and experimental metapopulations. Such an outcome would demonstrate the ineffectiveness of Phase III migration to add anything of consequence to local individual selection. However, the evolution of a significantly higher mean fitness in our experimental metapopulations than in the controls would confirm Wright's theory and illustrate the evolutionary efficacy of Phase III migration. We were eager to perform this direct experimental test.

In order to impose random migration in a control metapopulation at a level exactly equal to that of an experimental metapopulation, we first counted the total number of migrants, M_{total}, in the experimental migrant pool. We moved exactly the same number of migrants, M_{total}, among demes *at random* in the control metapopulation. After the census of demes in the control metapopulation, we drew demes at random and with replacement M_{total} times, and each time a deme was drawn, we added a migrant from it to the pool of migrants. This gave us M_{total} migrants, each drawn at random from among the 50 control demes. How should we distribute migrants out of this pool and into control demes again *at random*? Again, we drew demes at random and with replacement M_{total} times, and for each time a deme was drawn, it received a migrant from the pool. In this way, we randomly moved M_{total} migrants out of some demes and into others across the control metapopulation (Figure 9.4). Although the control demes differed from one another in W, just like our experimental demes did, with this protocol there could be no Phase III group selection in the control metapopulation. On average, high fitness and low fitness control demes contributed the same number of adult beetles to the migrant pool and also, on average, each received the same number of migrants from it.

With these protocols (Figures 9.3 and 9.4), we would move exactly the same numbers of migrants among populations in our control metapopulations as we did in our experimental metapopulations. The only difference between the experimental and control metapopulations was the source of the migrants: random in the controls but biased toward those demes with higher W in our experimentals. If the controls changed in average productivity, it had to be the result of selection *within* populations as standard

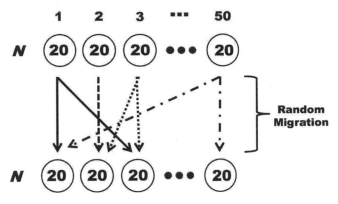

FIGURE 9.4 A schematic diagram of our protocol for a control metapopulation in our test of Wright's SBT.

theory predicted; it could not be owing to Phase III interdemic selection. If we observed the opposite outcome, where the experimentals changed in average productivity but not the controls, it would have to be the result of Phase III interdemic selection *among* populations acting as Wright's theory predicted.

Replication of the Experimental Metapopulations

Now that we knew *how* to impose Phase III interdemic selection by differential dispersion, we had to decide *how often* to do it. Practical considerations limited our choices. On the one hand, we could impose Phase III migration every generation in each of three replicate experimental metapopulations, each with its own control (a total of 300 populations). This replication would allow us to see how much variation there was from metapopulation to metapopulation in the efficacy of Wright's process. On the other hand, we were impressed by the McCauley-Wade finding that group selection only once in a while produced as much evolutionary change as group selection every generation. On this basis, we thought we should vary how often we imposed Phase III migration. We could do it every generation (E1), every two generations (E2), or once every three generations (E3). And, each experimental metapopulation could have its own control metapopulation with the appropriate level and frequency of random migration (C1, C2, and C3, respectively). This would still be 300 total populations, but we would be sacrificing replication of any one of the experimental treatments. We did not believe we could handle more than 300 populations because each of us was already running other large experiments!

After much discussion, Charles and I decided on the E1, E2, and E3 metapopulations because they did provide "replicates" of Wright's Phase III group selection process, whose efficacy we were testing. And, given the McCauley/Wade results, we felt we had a real opportunity to see a measurable response to interdemic selection even in the E3 metapopulation, with the weakest Phase III migration. If our intuition was borne out, this would be the strongest evidence that Wright's process was effective because E3 Phase III migration was not only much weaker than in E1 but also it was intermittent. It would be a much stronger result than replicating E1 three times. In fact, we believed that either the entire experiment would fail or the entire experiment would succeed, no matter how we distributed our efforts across treatments. We decided that, if we were going to kill ourselves with more work, replicating the Phase III process with metapopulations E1, E2, and E3 was a much more interesting way to go than doing E1 three times.

For the initial generation, we used the c-SM laboratory strain to set up three metapopulations, each with 50 demes, each deme with 20 adult beetles (3 metapopulations × 50 demes × 20 adults per deme = 3,000 initial adults). We reared them under standard conditions and censused each deme two months later. After the census, we took one of the three metapopulations and made it into the separate E1 and C1 metapopulations by making *two* new, twin demes from each *one* of the original demes. To start the second generation, one twin was assigned to the experimental metapopulation, E1, and the other was assigned to the control metapopulation, C1 (see Figures 9.3 and 9.4 above). In this way, at the start of our experiment, each deme in an experimental metapopulation had a corresponding twin deme in its control metapopulation. As a result, each E and C metapopulation pair enjoyed a totally even start, genetically and ecologically speaking!

The Response to Phase III Interdemic Selection

With our protocols, the E1 metapopulation experienced the strongest group selection, followed by E2 and then by E3, the weakest of them all. Each could be compared to its twin control to see whether or not Phase III migration was causing an increase in average W. We also wanted to compare E1, E2, and E3 with one another. To do so, we required a method for standardizing the strength of group selection so that we could say *how much weaker* group selection was in E3 than in E1. The variation in the frequency of Phase III migration, every generation in E1 but only once in three generations in E3, meant that comparisons among them scaled by generation had little meaning.

We borrowed a technique developed by animal breeders for comparing responses among different, artificial individual-selection experiments. To make such comparisons, breeders used a standardized measure of *response to selection*, $R(t)$, and a standardized measure of the *strength of selection*, $S(t)$. (The "unit" of standardization in this case is the "standard deviation." It is the "sigma" in the well-known business strategy "Six Sigma," developed by Motorola.) Breeders would regress the response to selection after t generations, $R(t)$, on the total or cumulative amount of selection imposed up to that time. The total amount of selection imposed equals the *cumulative selection differential*, $C(t-1)$; it is the sum of $S(1)$, $S(2)$, $S(3)$, up through $S(t-1)$. Moreover, the regression of $R(t)$ on $C(t-1)$ is the classical definition of *realized heritability* (Falconer 1989; Wade 1982c; Wade and McCauley 1984; Hill and Caballero 1992; Wade and Griesemer 1998). As we saw in earlier chapters, in colloquial terms, realized heritability measures what you get (R) as a return on how hard you try (C). We were not trying as hard to change W in E3 because we were imposing Phase III group selection only once every three generations with no group selection in the other two.

It is easy to imagine how the differences in frequency of group selection might affect $C(t)$, the cumulative selection differential. In the E1 metapopulation, the interdemic selection differential at generation t, $S(t)$, was never 0, because we imposed group selection each and every generation. In contrast, during the odd-numbered generations (i.e., $S(1)$, $S(3)$, $S(5)$, etc.), $S(t)$ was always 0 in the E2 metapopulation because we did not impose any group selection in those generations. Similarly, in the E3 metapopulation, $S(1)$ and $S(2)$, $S(4)$ and $S(5)$, etc., were always 0, because we imposed group selection once in every three generations (see Table 9.1, next section). Thus, at any generation, $C(t-1)$ was largest for E1 and smallest for E3. These standardized measures allowed us to quantify how much stronger group selection was in E1 than in E2 or E3.

In individual selection experiments, the response to selection, $R(t)$ tended to be proportional to the cumulative selection differential, $C(1-t)$. This was expected, because, with a constant value of heritability, h^2, weaker selection—that is, smaller $C(1-t)$, results in a weaker response (Hill and Caballero 1992). In fact, the expected response is the product of h^2 times $C(1-t)$. We had a number of reasons, founded on past experimental results, for believing that the response to group selection might be different than the response to individual selection. First, g^2 was not a constant like h^2. We had seen g^2 increase with generation in response to metapopulation structure. For this reason, we believed that g^2 might be larger in E3 than it was in E1, because E3 had a lower migration rate; our earlier migration experiments had shown that g^2 was inversely proportional to migration rate (Wade 1982c; Wade and McCauley 1984). Second, our earlier results had shown that the response to group selection could be as large for a small selection differential (a single episode of group selection) as it was for a much larger, cumulative selection differential (many sequential episodes of group selection added together). Thus, we believed the response to interdemic selection might well be larger in E2 or E3 than in E1. One of our experimental questions was whether the response to group selection would be proportional to the strength of group selection as measured by $C(t)$, the cumulative selection differential.

The general expectation from standard theory was that our experiment would fail. It was thought that Wright's Phase III differential migration could not be significantly better than random migration for spreading good gene combinations among demes in a metapopulation. And, if we did observe a response, it would be much more likely to occur in the E1 treatment, where interdemic selection was stronger, than in either the E2 or E3 treatments, where it was weaker. Under the standard view, it was simply impossible that the response to group selection in E2 or E3 could equal or exceed the response of E1.

Results of the Shifting Balance Experiment

Just as the variations in cannibalism had created among-group selection in Wade (1980a; see chapter 7), here, the variations in productivity of the beetles determined the level of Phase III differential migration and thus the strength of interdemic selection in each of our three experimental metapopulations. The beetle-created variation in W determined the total numbers of migrants, M_{total}, in an experimental metapopulation and in its corresponding control metapopulation. By our protocols, the migrants in the experimental metapopulation came from the most productive demes and went into the least productive demes. The same number of migrants moved at random, out of and into demes in the control metapopulation.

Relative to our earlier studies of group selection by differential extinction, the strength of Phase III group selection by differential migration here was much weaker. The group selection differentials, $S(t)$, ranged from one tenth to two tenths of a standard deviation of the variation in W (Table 9.1, second to last row, S values lie between 0.107 and 0.234; Wade and Goodnight 1991). The average migration rate per deme per generation, resulting from the beetle-created differences in productivity, ranged from a low of 4.3% in the E3 and C3 metapopulations to a high of 9.5% in the E1 and C1 metapopulations. Notably, all migration values were greater than that previously considered to prohibit group selection in theoretical discussion. That is, all exceed 2.5%, the level of migration corresponding to "one migrant per population every other generation" for an N of 20, considered the

TABLE 9.1 The generation, group selection differential (S), average number of migrants per deme (Nm), and cumulative selection differential ($C[24]$) for the three experimental metapopulations of Wade and Goodnight (1991).

Generation	E1		E2		E3	
	S	Nm	S	Nm	S	Nm
1	0.160	1.34	0	0	0	0
2	0.215	1.78	0.187	1.58	0	0
3	0.168	1.40	0	0	0.224	1.74
4	0.241	1.98	0.234	1.94	0	0
5	0.177	1.54	0	0	0	0
6	0.406	3.34	0.318	2.62	0.409	3.22
7	0.262	2.06	0	0	0	0
8	0.201	1.79	0.300	2.30	0	0
9	0.146	1.20	0	0	0.249	2.12
10	0.188	1.46	0.177	1.88	0	0
11	0.444	3.40	0	0	0	0
12	0.240	1.86	0.350	2.70	0.477	3.80
13	0.246	1.86	0	0	0	0
14	0.237	1.98	0.265	2.10	0	0
15	0.445	3.60	0	0	0.643	5.34
16	0.178	1.32	0.299	2.24	0	0
17	0.287	2.32	0	0	0	0
18	0.086	1.20	0.255	2.32	0.137	1.08
19	0.245	1.80	*0.219*	1.48	0	0
20	0.231	1.64	0	0	0	0
21	0.184	1.56	0.087	1.60	0.259	2.00
22	0.210	1.60	0	0	0	0
23	0.262	2.24	0.135	1.04	0	0
24	0.145	1.28	0	0	0.167	1.32
Average	0.234	1.90	0.118	0.99	0.107	0.86
Sum, C(24)	5.064	45.55	2.826	23.80	2.565	20.62

upper limit for group selection (see chapters 3 and 4). Our levels of migration were also larger than $(1/2N)$, which was approximately 0.027 (discounting N adults by 0.9 to obtain N_e as per chapter 4); so migration in our experiment was stronger than drift. As a result, according to standard theory, our metapopulations were not significantly subdivided; instead, they were rather close to large randomly mating populations. The migration among demes in the control metapopulations was more than sufficient to test the prediction of standard theory that random migration would spread favorable gene combinations (if they existed) and that Phase III was unnecessary.

Over twenty-four generations, $C(24)$, the cumulative strength of group selection in E1 was 80% greater than it was in E2 (5.064 versus 2.826, respectively; see $C[24]$ last row of Table 9.1) and 97% greater than the $C(24)$ of E3. The cumulative group selection differentials were comparable for E2 and E3: 2.826 and 2.565, respectively (see $C[24]$ last row of Table 9.1). This occurred because there was more variation in W among the E3 populations than there was among the E2 populations when we imposed Phase III group selection. (This can be seen in their respective $S[t]$ values in Table 9.1, where the non-zero values of S for E3 are often greater than those of E2.)

Result 1

Phase III group selection significantly increased mean fitness, W_{mean}, in each of the three experimental metapopulations relative to that of the corresponding control with the same level of random migration. This finding was the first empirical confirmation of the efficacy of Phase III of Wright's Shifting Balance Theory for causing a change in mean population fitness. Random migration in the control metapopulations did *not* have an effect equivalent to the Phase III migration of the experimental metapopulations, directly contradicting theoretical predictions (Barton and Rouhani 1993; Barton 1992). Whatever the latent genetic variation for fitness in an E metapopulation, that same variation was necessarily also latent in its paired C metapopulation, because, at the start of our experiment, each of the 50 populations in E1, E2, and E3 had a corresponding twin in C1, C2, and C3, respectively. Furthermore, the average level of control random migration was identical to the level of Phase III migration in its paired experimental metapopulation. Clearly, variation for fitness existed within our metapopulations; that variation was heritable; and, group selection by Wright's Phase III migration used that heritable variation to change mean fitness but individual selection with random migration had not.

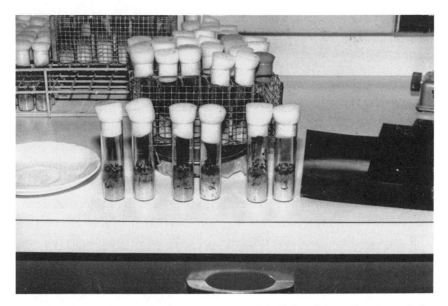

FIGURE 9.5 Six populations taken from generation 12 of the Shifting Balance Experiment. Each pair consists of one population from E2 (left) and its initial twin from C2 (right). The leftmost pair is population 1 from E2 and next to it is population 1 from C2; the next pair are populations E2-2 and C2-2, and the right-most pair are populations E2-3 and C2-3. The dark spots in the layer above the whiter flour are adult beetles. Notice that the left member of each population pair has many more beetles and much less flour than its twin from the control to its right. This difference in mean population fitness, visible without statistics, is the result of only six episodes of Phase III migration in the E2 metapopulation and six episodes of random migration of the same amount in the C2 metapopulation.

To see the equality of the levels of migration consider migration in the E2 and C2 metapopulations (Table 9.1, last row). By generation 24, there had been 24,000 breeding adults in each of these two metapopulations; and, each had had a total of 1,190 migrants moved by us among its demes. That is a total of 23.8 migrants per deme or an average of one migrant per deme per generation, twice the theoretical threshold believed to prohibit group selection. Yet, by generation 25, on average, the 20 founding adults of an E2 population produced 30 more adult offspring in 60 days than did 20 adults from a C2 population (204.3 mean for E2 versus 173.9 mean for C2).

The difference in number of adult offspring produced was visible without statistics as seen in a photo taken at generation 12, the midpoint of the Wade-Goodnight experiment, after only 6 episodes of Phase III migration (Figure 9.5). The photo shows population pairs 1, 2, and 3 from the metapopulations E2 (on left) and C2 (on right). (Remember, both populations in a pair were twins from the same single population at generation 1.) These data not only indicate the efficacy of Wright's Phase III migration as a means for increasing mean fitness, but also refute the theoretical argument that random migration alone is sufficient to distribute a local genetic advantage arising in one population to other populations across a metapopulation.

Result 2

The response to interdemic selection was not proportional to its strength: group selection in E2 was only half as strong as it was in E1, but the response in E2 was 2.5 times greater than the response in E1. With individual selection, the short-term response to selection increases with the strength of selection. Similarly, the response is greater for strong selection than it is for weak selection (Hill and Caballero 1992). The response to our Phase III group selection was strikingly different. The cumulative selection differentials, $C(24)$, in the last row of Table 9.1 measure the strength of interdemic selection. The value of $C(24)$ for E1 is 5.064, while the value for E2 is 56% smaller, 2.826. That is, Phase III group selection was 56% weaker in E2 than it was in E1 ($[2.826/5.064]$ = 0.56). Nevertheless, the response to group selection was 268% greater in E2 than it was in E1. Much weaker group selection but a much stronger selection response is a result that is simply impossible in standard theory with only additive gene effects. It is as though two people, equally fast, run to a store ten blocks away; but, one runs every other block walking the blocks in between, while the other runs every single block. In an additive world, the person who runs every block arrives at the store first, before the person who runs only every other block. The response of E2 relative to E1 was the equivalent of the second person arriving at the store first. Although surprising by the expectations of standard theory, this result was congruent with the earlier McCauley-Wade results, which had motivated our decision to set up separate E1, E2, and E3 metapopulations instead of setting up 3 replicates of E1. In fact, we could not have observed this result if we had set up 3 replicates of E1 and C1.

The nonproportional response is best illustrated with the *realized group heritability*, the response relative to S. The realized group heritability was 0.207 (SE ± 0.043, P < 0.001) in the E1 metapopulation; 0.554 (SE ± 0.229, P < 0.015) in the E2 metapopulation; and, 0.205 (SE ± 0.050, P < 0.001) in the E3 metapopulation (Wade and Goodnight 1991). These realized heritabilities were significantly different from one another (P < 0.050)—under the standard theory, they should have been the same. Remarkably, the realized heritability was more than twice as large in E2 as it was in either E1 or E3.

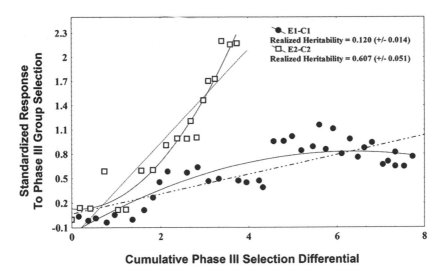

FIGURE 9.6 The standardized response to Phase III group selection plotted against the cumulative selection differential (C[t] defined in text) for the E1-C1 and E2-C2 metapopulation pairs after seven years of selection. Despite twice as large a C(t) for the E1-C1 pair, the response to Phase III interdemic selection was not only much greater in the E2-C2 pair, but also accelerating.

The differences Charles and I reported in 1991 became even larger as I continued Phase III selection (see Figure 9.6). After seven years, the response to Phase III group selection was *five times* greater in E2 than in E1. Moreover, the response in E1 appeared to have plateaued or leveled off, with population mean fitness in E1 approximately 0.70 standard deviations greater than that of its control, C1. In sharp contrast, the difference in W between E2 and C2 was three-fold larger, exceeding 2.00 standard deviations, and still increasing when I finally stopped the experiment in 1995. It is important to emphasize that Phase III selection remained **twice** as strong in E1 as it was in E2, 8 standard deviations versus 4 (see x-axis of Figure 9.6). The response in E1 was much smaller than in E2, however, because g^2_{E1} was smaller than g^2_{E2}. The difference in realized response appeared to accelerate as shown by a quadratic regression, which fit the data significantly better than a linear model (quadratic model accounted for 88% of the variance of realized g^2_{E2}).

Result 3

The variance in migration rate was greater for Phase III migration than it was for random migration. As a result, the **effective migration rate** *was lower in* *our experimental metapopulations than it was in the corresponding controls.* I first noticed this somewhat counterintuitive result when I tried to use the change in F_{ST} from one generation to the next as a means for estimating N_e. Although I had definitively measured N_e in earlier studies (Wade 1980b, 1984c), I had not measured it during a group selection experiment. I believed the significance of our findings for Wright's SBT would be enhanced if I could estimate N_e in the midst of our experiment.

I measured F_{ST}, from observing the variance among populations in the frequency of the black body color mutation (b), which was segregating in our metapopulations. I scored all individuals for genotype (b/b, $b/+$, or $+/+$) in the C2 metapopulation at consecutive generations 13 and 14. At generation 13, the frequency of b was 0.0625 (SE ± 0.010) with an F_{ST} of 0.029 and, at generation 14, it was 0.0754 (SE ± 0.014) with an F_{ST} of 0.047; statistically, the frequencies were the same. Furthermore, in both generations the frequency of b was not correlated with W. That is, the frequency of b did not affect the productivity of a group of 20 adults.

I inserted these F_{ST} values into Wright's equation relating F_{ST} at one generation to F_{ST} in the next:

$$F_{ST}(t+1) = \{(1/2N_e) + ([1-(1/2N_e)][F_{ST}(t)])\}$$

I then solved for N_e. In good agreement with our two earlier estimates (Wade 1980b, 1984c), the ratio of (N_e/N) was 0.90. This meant that, viewed through the lens of the b allele, our beetles were behaving very nearly like the idealized genetic entities of the standard theory during our shifting balance experiment. The significance of this finding was that while single gene frequencies were behaving in accord with the expectations of standard evolutionary theory, the ecological interactions among beetles influenced by those genes were not.

When I extended this study of F_{ST} to the E2 metapopulation, I noticed that it had a larger value of F_{ST} than that observed in the C2 controls (Wade 2013). At first I thought I had made a mistake, and I re-checked the numbers of migrants to make sure they were identical in both metapopulations. They were. Thinking further, I realized that Phase III migration results in *greater variation in migration* among demes than does random migration. This happens because the most productive demes in our E metapopulations receive no migrants; they only send out migrants. Moreover, when g^2 is positive, the most productive demes in one generation tend to be the most productive demes at the next. This meant that they might not receive migrants for two consecutive generations (or more!). In contrast, with random migration at the levels we imposed, it was highly unlikely that any deme would go without migrants two or more generations in a row. As a result, the *variation* in migration from one deme to the next was definitely greater in an E metapopulation than it was in the corresponding Control, even though the mean migration rates were identical.

Such variation meant that migration was less effective at mixing together the genetic differences among populations in the E metapopulations. Theory shows that variation greater than random reduces the *effective migration rate* (Sved and Latter 1977; Whitlock 1992). And, with a high g^2 (Wade and Griesemer 1998), high fitness populations were immunized against migration; that is, they were sources but not sinks for migrants. Thus, another consequence of g^2 in the SBT was to *increase* the variability in migration from one population to another over random (Wade 1996, 2013). Here was a source of variation right under my nose that I did not anticipate despite 15 years of intensely studying and testing Wright's work. Of course, neither Wright nor his critics had noticed it either.

How much larger was the variance in migration rate? I computed the total number of migrants per deme in the three experimental and three control metapopulations (50 demes per treatment; cf. Wade and Goodnight 1991) for generations one through 13 from our census data and experimental records. Table 9.2 lists the total number of migrants, the mean number

TABLE 9.2 The total number of migrants moving among demes in each control and experimental pair of metapopulations; the average number of migrants per deme (M) and the variance among demes in the number of migrants received (V); and (V_E/V_C), the ratio of the variance among demes in migrant numbers of the experimental to control metapopulations (Wade 2013).

Metapopulation	Total Migrants	M	V	V_E/V_C
C1	1,244	24.88	24.35	4.3
E1	1,239	24.78	104.71	
C2	650	13.00	14.20	3.2
E2	636	12.72	44.94	
C3	544	10.88	14.15	4.8
E3	543	10.86	67.55	

of migrants per population (M), the variance in number of migrants among populations (V_M), and the ratio (V_E/V_C) of the variance of migration of each experimental to its control metapopulation. It is not surprising that the total number of migrants is a function of the periodicity with which migration was imposed. Imposing migration every generation (E1, C1) resulted in more total migrants than imposing migration every two (E2, C2) or every three (E3, C3) generations (Table 9.2 column 2). (The small differences in the E and C totals represent minor errors in executing our protocols over 2.5 years.) The totals and the means per population show that each pair of control and experimental populations experienced the *same average* migration rate (Table 9.2, columns 2 and 3). However, the last two columns of Table 9.2 reveal that the variance in the total number of migrants was 3- to 5-fold higher with Phase III differential migration than it was with random or island model migration (variance ratio test, $P < 0.001$ in each case).

Why is this increase in the variance in migration under Phase III and its effect on F_{ST} important? *It is important because it means that Phase III interdemic selection is both adaptive and diversifying at the same time.* That is, by making F_{ST} larger, the frequency of *b* varied more among the group-selected demes than it did among the control demes. A greater genetic variance among demes makes future interdemic selection on *b* or any traits it affects easier. Differently put, the greater variation in the frequency of the *b* allele among E2 demes than among C2 demes means that the *b* allele frequency has a higher g^2 in E2 than it does in C2. As a result, it would be easier to increase (or decrease) its frequency by group selection in the E2 metapopulation than it would be in the C2 metapopulation. Remember, for a single gene like *b*, F_{ST} and g^2 are the same.

The implication of this finding is that interdemic selection acting on one trait, mean fitness, created even more favorable conditions for future interdemic selection acting on another, uncorrelated trait, the frequency of the *b* allele. A long-standing criticism of Wright's SBT is that it requires low levels of migration at Phase I in order for random drift to increase g^2 but then high levels of migration at Phase III to effectively export good gene combinations (e.g., Coyne et al. 1997) and these high levels of migration reduce or eliminate g^2. Our finding that Phase III interdemic selection increases g^2 obviates this criticism.

Individual selection was not like this. High intensity individual selection produces greater short-term responses but smaller long-term responses (López-Fanjul 1989; Hill and Caballero 1992). Individual selection acting to spread a gene with a positive effect on fitness is slowed by other genes with *either* positive or negative effects on fitness (Fisher 1930). That is, simultaneous individual selection on more than one gene is slower than selection on either gene separately. In fact, strong individual selection at one gene is experienced by all other genes (especially those closely linked to it) as an increase in the strength of random genetic drift, with a consequent loss of genetic variation at all other genes (Hill and Robertson 1966; Barton 1995). Hence, *strong individual selection diminishes its own future efficacy. The increase in F_{ST} with Phase III migration shows that strong interdemic selection does the opposite; it enhances its future efficacy.*

Beyond the Shifting Balance

Each of our three findings was novel and surprising. First, our findings refuted the prediction of standard theory that random migration was sufficient to spread good gene combinations and that Phase III migration was unnecessary. Second, the several-fold greater response to less frequent interdemic selection was not only unexpected but also not possible under standard theory (Figure 9.6). Longer-term individual selection tends to produce more accurate and less variable estimates of realized h^2 (Hill and Caballero 1992). In our case, longer-term group selection produced estimates of realized g^2 between the E2 and E1 metapopulations more different in 1995 than our published estimates in 1991 (Wade and Goodnight 1991). In fact, where the response of E1 above C1 had plateaued, in E2 relative to C2 it had accelerated. Third, Phase III migration was *less* effective genetic mixing than random migration at the same rate. This meant that g^2 was not as strongly influenced by Phase III differential migration as it was by random migration. Critics of Wright's SBT (e.g., Coyne et al. 1997) considered Phase I and Phase III to be in conflict with one another. Not only had we shown that *infrequent* Phase III migration was more effective at changing mean fitness than frequent Phase III migration, we had shown that the variation among demes associated with Phase III migration preserved g^2 more effectively than did random migration.

Like my doctoral research, our results raised as many questions as they answered. Why did our results lie so far afield from the predictions of standard theory? Why was Phase III differential migration so much more effective than random migration and individual selection at changing W? Why was Phase III interdemic selection every two generations more effective than interdemic selection imposed more or less frequently than that? Could the response to interdemic selection have been even larger if the migrants from a particular source deme had dispersed together as a single cluster into other demes, instead of being mixed with migrants from other source demes? How many migrants did it take to convert a low W deme into a high W deme by Phase III migration? Did the key to our findings lie in the nature of the genetic basis of the response to interdemic selection?

I will report our experiments to investigate these and related questions in chapter 10. But, before moving on, I would like to acknowledge that the reception of our findings was mixed and they have had only modest impact on the discussion of group relative to individual selection. Some appeared to accept the power of group selection at the level of families even as they dismissed our findings as being irrelevant to Wright's SBT because "the scheme is a variant of the family-selection method used to select for quantitative traits of low heritability. . . . Such a design could successfully alter nearly any character, including bristle number in *Drosophila*, a classical subject of mass-selection experiments" (Coyne et al. 1997, p. 663). As discussed in chapter 5, throughout our experimental investigations of group selection, we freely borrowed concepts and methods from the animal and plant breeding literature, especially examples of family selection. Where our early studies were criticized on the grounds of "infringement" on genetic concepts like family selection, our later studies were dismissed on the grounds that our work was a variant of family selection. Ours was the first test of Wright's process to use an epistatic trait, total fitness, and we had shown empirically that the demes in our metapopulations were much larger than families (chapters 3 and 4). Moreover, we had studied family selection itself in the lab, in the field, and in theory for more than a decade (chapters 7 and 8), arguing throughout that family selection was group selection and demonstrating in field, lab, and mathematical model how and why it was better understood as such. Like criticism of my doctoral work, many maintained that Wright's process would not work. When experiments showed that it did, the argument shifted: it worked, but was unimportant.

The bias toward the standard view was strong and ingrained. Our carefully designed, replicated experiments had produced results not predicted by, contrary to, and, in some cases, unfathomable under standard theory. Moreover, they rested on the foundation of decades of prior laboratory, field, and theoretical results. Yet, they hardly moved the needle of opinion.

10 Beyond the Shifting Balancing Theory

Administrative Attention Deficit Disorder

Coincident with the publication of our SBT paper in *Science* (Wade and Goodnight 1991) I was appointed chair of the Department of Ecology and Evolution (E & E) at Chicago. At first this seemed like a promising vantage point from which to influence the shape of my discipline at Chicago and to work with other departments in neighboring fields, like geophysical sciences, biological anthropology, statistics, and human development, to recruit faculty and interdisciplinary graduate students. I was a loyal fan of our research faculty; we had one of the top graduate training programs in the country; and I had ideas to make our graduate programs better. Moreover, I knew and enjoyed working with the staff.

For a while, things went very well. Under the previous chair, Dr. Brian Charlesworth, we had hired two prominent molecular evolutionary geneticists, Drs. M. Kreitman and C.-I. Wu, who arrived just as I began my appointment. One of my tasks was to rebuild the other half of E & E, Ecology, with the dean allowing us 3–5 additional hires in this area. I was director of an NIH Program Project grant in evolutionary genetics, which was awarded after a reverse site visit to the NIH Bethesda campus. That award increased our funding for graduate and postdoctoral students and provided support for the department's *Drosophila* media kitchen, deionized water system, and several staff salaries. In addition, it brought in sufficient indirect costs to persuade the University to replace the ancient air-handling unit on the Zoology Building.

In Ecology, we made two foundational hires in Drs. M. Leibold and E. Simms. They in turn helped us recruit Drs. T. Wootton and K. Pfister. Together with our faculty in behavioral ecology, this young group of scientists grew into one of the strongest ecological research groups in the country.

Faculty from E & E and the Committee on Evolutionary Biology (CEB), chaired by Dr. Jeanne Altmann, also wrote two successful training grant applications. One was an NSF Biodiversity Training Grant and the other was a Department of Education, Graduate Assistance in Areas of National Needs Training Grant. In addition, we organized a course for graduate students of both programs to submit applications for pre-doctoral awards. This resulted in two to four pre-doctoral awards per year from the NSF, the Howard Hughes Medical Institute, and the Environmental Protection Agency—it was as though we had a third training grant!

With a bit of social engineering, Jeanne and I reduced the average time to a doctoral degree from 6+ years to just over five years. Time-to-degree had such a strong effect on the graduate program rankings by the National Research Council that, in my second term as chair, we had moved our national ranking from 20th to first, where we were tied with Stanford. Unfortunately, the elite national ranking proved to be a double-edged sword. For graduate training, it was an unqualified good, with the size of our applicant pool tripling during this period. On the faculty side, the effects were mixed. There was conflict over who was "most" responsible for the program-wide recognition. This myopic conflict completely ignored the contributions of CEB faculty outside the department and, indeed, outside the university, in the Field Museum, the Brookfield Zoo, and the Shed Aquarium, which were essential to our reputation. The ranking also precipitated an increase in outside offers to faculty, which further destabilized the cohesiveness of our group. To those outside the program, Chicago appeared to be doing great, but, on the inside, we were struggling.

After the Shifting Balance Experiment

The Genetics of Population Fitness

Our finding that interdemic selection every other generation had been most effective for increasing mean fitness was novel, especially in light of expectations based on standard theory. We needed to understand why. Results from artificial individual selection experiments gave little guidance. More frequent individual selection was always more effective than less frequent selection because the heritability (h^2), on which the selection response depends, remained essentially constant from one generation to the next. Unlike the constancy of h^2, our experiments had shown repeatedly that g^2 tended to increase over time. We believed that our response to group selection involved a complex interaction between random drift, group selection, and g^2.

We knew that random drift created g^2 and that group selection reduced it (at least for the target trait, if not for other traits). Our hypothesis was that the optimal balance between the increments to g^2 from drift and the decrements from group selection occurred when we imposed group selection every two generations. If we imposed selection more often (every generation as in E1) drift did not create sufficient g^2 to sustain a strong selection response. If we imposed selection less often (as in E3), we were not efficiently using the g^2 created by drift. Our hypothesis was straightforward, but it was mechanistically insufficient since it was silent on the underlying genetic mechanism. We did not have a genetic model of either indirect genetic effects or gene interactions that could act as required by our hypothesis.

In Phase III of the SBT, migration out from demes of high mean fitness and into demes of lower mean fitness causes an evolutionary increase in average fitness across a metapopulation. Wright called the genetic transformation of a deme from low fitness to high fitness a *peak shift* (Wright 1931, 1969). Both differential migration and local individual selection played a role in the process of a peak shift, but Wright did not discuss the relative magnitudes of each role. The theoretical argument that Wright's Phase III was unnecessary for a peak shift assigned the largest role to individual selection. However, if genes with indirect effects were the cause of peak shifts, as we believed, individual selection could play only a subsidiary role if any at all.

We also knew that there were several features of Nature that Wright had not considered. Biotic and abiotic spatial variations result in genetic tradeoffs between different local adaptations. As a result, there was no necessary correlation between adaptive function at the scale of local demes and function at the global scale of the entire metapopulation. Would a gene combination that was *locally adaptive* confer high fitness in another deme in another environment? Empirical evidence for genotype-byenvironment interactions (G × E) from our lab (Wade 1985a, 1990) and those of many others (Via 1984 a and b; Shaw 1986) demonstrated that high fitness in one deme could mean lowered fitness in another. Whenever this was the case, Phase III migration, instead of initiating a shift to a higher fitness peak, might lower mean fitness in a deme receiving migrants—group selection in the opposite direction to Wright's process! In our SBT experiment, we had studied only one-way migration, from high fitness into low fitness demes; we had not studied migration in the other direction. And, in many experiments, we had shown that, when group selection by differential extinction favored low fitness demes, mean fitness declined but interspecific competitive ability increased. Group selection in the laboratory worked well in both directions and there were plausible scenarios for its acting in both directions in Nature as well.

Using the McCauley and Wade (1980) study as a model, we chose populations from the C2 and C3 metapopulations to investigate whether or not migration from one deme into another had a measurable effect on mean fitness and, if so, how large and persistent that effect could be (Wade and Griesemer 1998). We were confident that demes from these two metapopulations were genetically different from one another despite periodic bouts of random migration among them in their history. Like McCauley and Wade (1980), we studied migration from demes of high fitness into demes of low fitness as well as the reciprocal: migration from low fitness demes into higher fitness demes. Unlike our earlier studies, we also studied migration between demes of comparable fitness (i.e., between two different high fitness demes or between two different low fitness demes). These crosses involved equal numbers of male and female adult beetles taken from different source or "parent" demes per the experimental design in Table 10.1. In addition, we tried to discover how much migration was necessary to cause a measurable effect on mean fitness by varying the amount of migration between pairs of demes (Table 10.2).

How productive would the "offspring" demes be if we crossed different numbers of males and females from different pairs of parent populations (Table 10.2)? Which parent, the more productive or the least productive, left a larger imprint on them? Was resemblance between a parent and its offspring demes proportional to the number of migrants that that parent contributed to the founding propagule of adults? We could answer these questions by regressing productivity of the offspring demes on the proportion of migrants contributed by each parent deme. This was the same parent-offspring regression method used by classical geneticists to estimate h^2 but here we applied it to parent-offspring demes to estimate g^2.

TABLE 10.1 The number of replicates of within- and among-deme crosses set up using 10 males from the deme in the top row crossed to 10 females of the deme in the left column.

	Deme 1	Deme 2	Deme 3	Deme 4	Deme 5
Deme 1	**6**	6	6	6	6
Deme 2	6	**6**	6	6	6
Deme 3	6	6	**6**	6	6
Deme 4	6	6	6	**6**	6
Deme 5	6	6	6	6	**6**

TABLE 10.2 The numbers of adults (equal number of males and females) from two different demes.

Parent	Deme i	0	4	8	12	16	20
Deme	Deme j	20	16	12	8	4	0

Results

First, we saw that the parent demes from both metapopulations (the bold diagonal populations in Table 10.1) were very different from one another in mean fitness, just as we had expected. For both metapopulations, the most productive parent deme produced twice as many offspring as the least productive deme (C-2: 65.0 ± 8.7 versus 146.0 ± 11.7; C-3: 104.0 ± 19.8 versus 189.7 ± 15.9). This confirmed that g^2 was not zero: there were significant genetic differences in mean fitness among demes in both metapopulations, despite random migration among them.

Our regression estimates of g^2 in our C2 and in our C3 metapopulations are shown in Figure 10.1. The values of g^2 were very high and essentially the same: 0.69 ± 0.124 in C2 and 0.641 ± 0.140 in C3. This regression of offspring on all parental pairs, however, hid many special cases in the average. Particular pairs of populations showed significant deviations from the linear regression (Wade and Griesemer 1998, p. 143). In some cases, adding a few migrants into a low deme from a high-producing deme increased offspring productivity much more than predicted by linear regression. And, conversely, in some cases, a few migrants from one parent into the other or vice versa had little or no effect on productivity; it took greater numbers of migrants before we could detect a measurable effect.

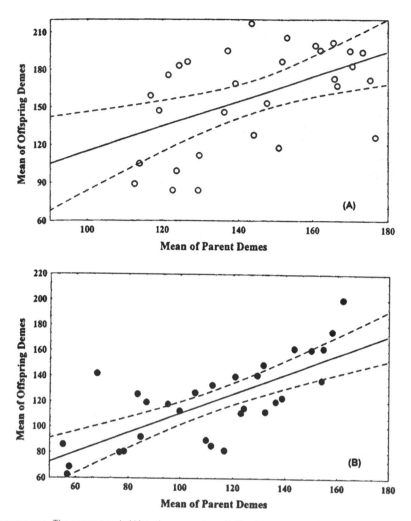

FIGURE 10.1 The upper graph (A) is the regression of offspring on parent fitness for the C2 metapopulation. The lower graph (B) is the same regression for the C3 metapopulation.

These studies showed convincingly that the C2 and C3 metapopulations harbored large amounts of heritable variation for fitness among demes. At the same time, there were some pairs of demes where the exchange of a few migrants had an inordinately large effect and, conversely, exchange of many migrants had an inordinately small effect. Nevertheless, in the control metapopulations mean fitness did not evolve to a higher value solely as a result of the random migration among demes followed by individual selection within them as claimed and predicted by standard theory. And, in both of these control metapopulations, g^2 was high, especially for a trait like total fitness, which should have little or no h^2 under standard theory.

Dominance and the Purging of Deleterious Recessive Alleles

My colleague Brian Charlesworth suggested a genetic mechanism that might depend upon just our hypothesized kind of interaction between drift and group selection: dominance (also known as "inter-allelic" epistasis). He hypothesized that interdemic selection in our SBT experiment had increased mean fitness by purging deleterious recessive genotypes from our metapopulations. Populations with high mean fitness, by chance, had few homozygous, recessive genotypes, while populations with low mean fitness had many. Under this hypothesis, our group selection protocol was a process for purging deleterious recessive genotypes from our experimental metapopulations. Because purging depends upon the frequency of homozygotes, the optimal regime for purging could have been interdemic selection every other generation. By chance, we had stumbled upon a selection schedule that allowed the frequency of homozygotes to increase by drift for two generations and then be removed by interdemic selection against those demes where homozygotes were most common. Under this hypothesis, less frequent group selection relative to drift (as in E3), allowed drift to fix deleterious alleles as homozygotes increased faster than interdemic selection could remove them. Conversely, more frequent group selection (as in E1) weakened the impact of drift on homozygosity and thereby diminished g^2. Most importantly, Brian's hypothesis could be tested.

Under this hypothesis, the large differences among our metapopulations in mean fitness were the result of differences in the degree to which they had been purged of deleterious recessives. If correct, when we imposed inbreeding, we should see differences between the E and C metapopulations in the rate of decline in mean fitness with inbreeding (also known as *inbreeding depression*). Specifically, we should find greater inbreeding depression in the demes of our control metapopulations than in our experimental demes, where the purging effects of group selection had occurred. Moreover, of all the experimental metapopulations, the smallest amount of inbreeding depression should occur in the E2 demes: they had the highest mean fitness presumably because they had experienced the most effective purging. This was a genetic hypothesis with an explicit connection to mean population fitness and it had the potential to explain our results. We had to test it.

To study the effects of inbreeding on mean fitness, I chose four demes at random from each of the six SBT metapopulations. From each of these 24 demes, we collected 12 virgin males and females and set them up as single mating pairs to reproduce (12 families per deme × 24 demes = 288 total families). Every 50 days, we censused the number of adult offspring produced by each pair, and selected a single male and female (1 brother and 1 sister) from among them to found the next generation. We carried on this brother-sister style of inbreeding for nine generations.

Inbreeding depression should have manifested itself in two ways. First, as time went by, some of our inbred lineages, the most inbred, should stop producing offspring and go extinct. Second, the surviving lineages should produce fewer and fewer offspring as they became progressively more inbred. If the purging hypothesis for our SBT findings was correct, we would expect the group-selected Experimental metapopulations to manifest less inbreeding depression than the unselected Control metapopulations. Specifically, we would expect more C families to go extinct than E families and we would expect steeper declines in the production of offspring by surviving C families than by E families. Moreover, the largest difference observed in inbreeding depression should be that between the E2 and C2 families. What did we find?

Results

We found inbreeding depression of both sorts, but the pattern of inbreeding depression was the same for the E and C families. In Figure 10.2, the fraction of surviving families from the E2 and C2 metapopulations are plotted as a function of generation. Notice that, for the first three generations of inbreeding, none of the families from either the E2 or C2 metapopulation went extinct. A small but insignificant difference appeared at generation 5 and became larger by generation 9, but remained statistically insignificant.

We observed a decline in the production of offspring by surviving families in both the C and E treatments (Figure 10.3). Clearly, average offspring production in the first generation *before inbreeding begins* is much higher in the E2 families than it is in the C2 families. As inbreeding progressed, productivity declined for both groups of families, but the decline was greater for the E2 than for the C2 surviving families. This was the opposite of the prediction of the purging hypothesis; but, again, even this difference was too small to be statistically significant.

The inbreeding effects on E1-C1 and E3-C3 families are not shown because they are comparable in every respect to the E2-C2 data shown above.

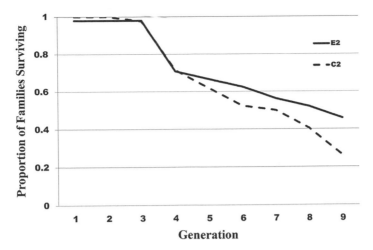

FIGURE 10.2 This graph illustrates the fraction of surviving E2 (solid line) and C2 (dashed line) families as inbreeding by brother-sister mating continued for nine generations. See text for further explanation.

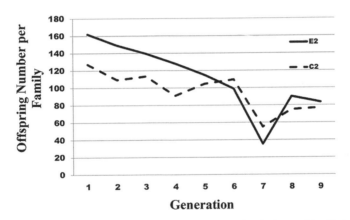

FIGURE 10.3 This graph illustrates the decline in offspring number in the surviving E2 (solid line) and C2 (dashed line) families as inbreeding by brother-sister mating continues for nine generations. See text for further explanation.

The results of our inbreeding study did not confirm the predictions of the genetic purging hypothesis.

The Genetics of Individual Fitness and Population Mean Fitness

One of our long term goals was establishing a direct connection between individual genetic variation for fitness (h^2) and among-population genetic

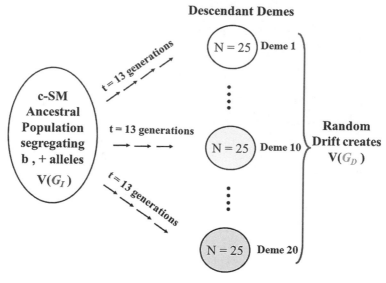

Descendant Demes

c-SM Ancestral Population segregating b , + alleles $V(G_I)$

$t = 13$ generations → → →

$t = 13$ generations → → →

$t = 13$ generations → → →

$N = 25$ Deme 1

$N = 25$ Deme 10

$N = 25$ Deme 20

Random Drift creates $V(G_D)$

FIGURE 10.4 A schematic diagram of the variation among demes, $V(G_D)$, in a metapopulation owing to random drift from the genetic variation among individuals, $V(G_I)$, within the ancestral population from which the demes were founded.

variation for mean fitness (g^2). The genetic differences among demes ($V[G_D]$ in Figure 10.4) had to originate from genetic differences among individuals. Moreover, under certain assumptions, an explicit formula connected the two levels of genetic variation. Up to this point, we had never directly estimated the genetic variation among individuals for fitness because it was technically difficult to keep track of the offspring of each individual beetle, male and female, in a population of beetles. I realized that, with the same black-body color mutation I was using to estimate F_{ST}, I could estimate the genetic variation, $V(G_I)$, for *individual* fitness. The numerator of h^2, $V(G_I)$, could be estimated using the standard paternal half-sib design of animal breeders (Figure 10.5).

Normally, with a design like that shown in Figure 10.5, one would collect offspring from each female and weigh or measure some physical trait on each of them. I wanted to measure their total fitness and I wanted to do it across a metapopulation of other c-SM beetles. So, I created the metapopulation shown in Figure 10.4 with adult beetles from the same c-SM stock population. Each generation, each deme was reduced to 25 adults and allowed to differentiate by random genetic drift without migration and without group selection. In the (N, m) parlance of our earlier work (Wade and McCauley 1984), it was a 25-0 metapopulation, and like all the earlier meta-

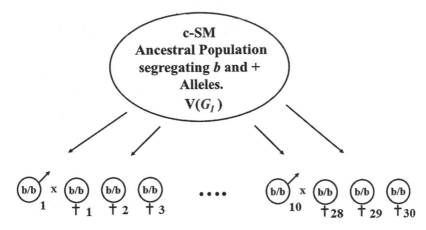

FIGURE 10.5 A schematic illustration of the standard half-sib design of quantitative genetics applied to the c-SM stock population. Virgin males and females are bred from the stock and each of 10 males is mated to 3 females. Subsequently, offspring from each female are measured. The variations in these offspring measurements can be used to estimate the genetic variation, $V(G_I)$, in the c-SM population, the amount of variation accounted for by genetic differences between the males (sires).

populations, g^2 increased over 13 generations from an initial value near zero to the very high value of 0.85. It was at this point that I combined the design for measuring $V(G_I)$ of Figure 10.5 with the metapopulation design measuring $V(G_D)$ of Figure 10.4.

To combine these two designs, I took two b/b males and two b/b females from each family in Figure 10.4 and dropped them individually into a group of 15 +/+ adults from one of the metapopulations in Figure 10.4 to create a new population (see Figure 10.6). Since there were 30 females and 10 populations, I did this 1,200 times, creating a very large number of beetles and demes — so many that the experimental populations had to be spread out across two incubators. The total number of b/b beetles was 1,200 (10 males × 3 females/male × 4 offspring/female/deme × 10 demes = 1,200 b/b adults) and the total number of demes, each with 1 b/b and 15 +/+ adults, was also 1,200. The experimental setup was similar to the one I had used to measure N_e (Wade 1980b, 1984c) except that we imposed a *genetic* structure on the b/b marked beetles (they were paternal half-siblings) and a *genetic* structure on the +/+ unmarked beetles.

The design depicted in Figure 10.6 allowed us measure $V(G_I)$ for individual fitness and $V(G_D)$ for population mean fitness, simultaneously. This experiment is designed to answer the question: which is more important in determining an individual's fitness, its own genes or the "social genetic con-

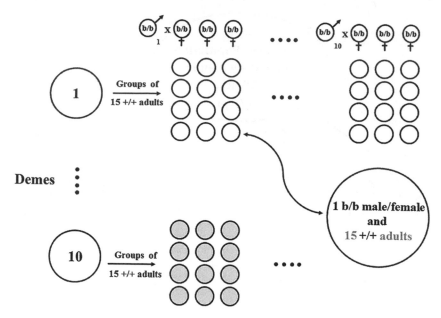

FIGURE 10.6 A schematic illustration of the experiment combining the standard half-sib design of quantitative genetics (depicted in Fig. 10.5) for estimating $V(G_I)$ with the standard McCauley-Wade metapopulation experiment for estimating $V(G_D)$ *(depicted in Fig. 10.4)*. The ten different demes represent ten different social contexts in which to measure the total fitness of a b/b individual of known pedigree.

text" of the population where it reproduces? If an individual's genes determine its fitness, then offspring from the same father should have similar fitnesses no matter which deme they live in. This would appear as column differences in Figure 10.6, where the identity of the father changes every three columns. Conversely, if the social genetic context determined an individual's fitness, then where a b/b beetle reproduced would matter more than its paternity, because offspring from different fathers reproducing in the same population should have similar fitnesses. This would appear as row differences, since the identity of the population changes with every four rows in Figure 10.6. Which is more important for individual fitness, column or row; genes or gene pool; individual or society?

Moreover, because the genetic differences among populations were derived from genetic differences among individuals, the estimates of $V(G_I)$ and $V(G_D)$ should be related to one another by a simple equation from standard theory:

$$V(G_D) = (2F_{ST})V(G_I).$$

We knew that with $N = 25$ and $m = 0$, that the value of F_{ST} after 13 generations was approximately 0.15, because we had estimated it many times before. This meant that $V(G_D)$ should be only 30% of $V(G_I)$. All we had to do was sort and count a myriad of +/b offspring of the b/b parent beetles, calculate $V(G_I)$ and $V(G_D)$ for fitness and see if the prediction of standard theory was true.

Implementing the design proved to be much more difficult than I expected. Most of the 1,200 demes produced a hundred or more offspring (Figure 10.7). For each deme, we had to sort and count the b/b individual (black), its +/b offspring (tan) and the +/+ founders (red) together with the +/+ offspring (also red) that the +/+ adults had produced by mating with other +/+ adults.

Results

We found that $V(G_D)$ did not equal 0.30 × $V(G_I)$ as predicted (Wade 1985a, 2000). Instead, $V(G_D)$ was a startling 18 times larger than $V(G_D)$! This was not possible under the standard theory. The prediction of standard theory had

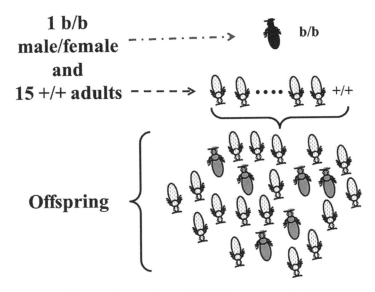

FIGURE 10.7 A schematic illustration of a population in the 1,200 experiment after the founders had produced offspring. The black beetle is the b/b founder individual; the gray beetles are the +/b offspring that the b/b founder produced by mating with one or more of the 15 +/+ founders; and the spotted beetles are the 15 +/+ founders and the +/+ offspring that they produced by mating with one another. The number of gray beetles in the population equals the total fitness of the b/b founder.

missed the mark by a factor of 60 (18/0.3 = 60). In short, social context (i.e., the population) mattered much, much more than paternity in determining individual fitness. This was clear evidence that a population of beetles was much more than the sum of its individual parts. And, it reinforced our experimental finding (Table 10.2) that migrants into some demes could have an inordinately large or an inordinately small effect on mean population fitness.

In fact, the offspring of one sire might have a high fitness in population 1 but a low fitness in population 5, while another sire might have a high fitness in population 5 and an intermediate fitness in population 1. That is, the sires changed ranks in the different populations (Wade 2000). Such changes were not possible with sum-of-the-parts theory; they could *only* happen with gene or genotype interactions (Goodnight 1988).

Questions Remain about the Genetic Basis of Response to Group Selection

Many further questions remained to be answered at the conclusion of these experiments. However, NIH reviews of my continuing proposal were not favorable. Moreover, reviewers indicated that continued funding for this line of research required that we develop tools for molecular evolutionary genetics and use them to map and identify the genes in *Tribolium* underlying fitness variation. I could understand the reviewers' reasoning, but in 1997, I did not have a wet lab necessary for such research at Chicago. Furthermore, molecular tools were not sufficiently developed for mapping genes in *Tribolium*. Indeed, the publication of the first assembly of the *Tribolium* genome was more than a decade away. At this same time, my beloved technician, Mrs. Ora Lee Lucas (Figure 10.8), encountered medical problems, which prohibited her from continuing to work in my lab. Mrs. Lucas had worked for Dr. Park for 17 years and worked, without interruption, in the Wade lab for another 21 years. She made our beetle work possible and I found it difficult to imagine how we could continue such a large research program without her.

I moved to Indiana University in the summer of 1998, where a small wet lab was a key part of my start-up package. With the help of many generous colleagues and graduate students at Indiana, particularly Dr. Paula Kover, the Wade lab began its new phase of molecular genetic studies, focused on the topic of speciation genetics. I saw speciation genetics as a means to continue studying gene interactions. Although gene interactions were

FIGURE 10.8 Mrs. Ora Lee Lucas, *Tribolium* technician extraordinaire, who worked in the University of Chicago laboratories of Dr. Thomas Park and Michael Wade from 1956 through 1997.

considered irrelevant in most of evolutionary genetic theory, they were believed to be essential in the origin of new species. Epistasis was critical to the evolutionary process whereby a single, interbreeding species split into two, distinct daughter species, each incapable of exchanging genes with the other. Genes that worked well together in the genetic background of a single daughter species were believed to work poorly when combined with genes from the other background in hybrids. The central paradigm of speciation genetic theory was the belief that changing the genetic background changed the fitness effect of a "speciation gene," from positive in the pure species background to negative in the hybrid background. Hence, the attitude of geneticists toward gene interactions varied depending on their research focus. Within species, most thought that the study of epistasis was futile and should be abandoned (Wade 2000, p. 215). Between species, epistasis was considered the key to unlocking Darwin's "mystery of mysteries," namely, the origin of new species and biodiversity.

In my first year, while waiting for lab renovations to be completed, Drs. Jason Wolf and Edmund "Butch" Brodie III and I completed work on the first book devoted wholly to gene interactions, the edited volume *Epistasis and the Evolutionary Process* (2000, Oxford University Press). Articles therein described in detail the data, the methods, and the open, unanswered questions regarding gene interactions as we understood them at that time. Despite significant advances in the study of gene interactions and in animal breeding methods since, many of those same questions remain unanswered today.

Open Questions

A central question motivating the work discussed throughout this monograph is this: Does natural selection assemble adaptive gene combinations one gene at a time operating on the average "additive" effects of single genes or does natural selection choose directly among different gene combinations—that is, "systems of interacting genes"? Where Wright's classical view of gene combinations refers explicitly to genes within the genome of a single individual, my meaning encompasses interactions of this sort but is much broader. It includes interactions between genes in different individuals, where those individuals can be members of the same species, as in our beetle studies, or they can be members of different species, as in our ongoing theoretical studies of host-symbiont coevolution (Drown et al. 2013; Drown and Wade 2014).

At the level of the individual phenotype, there remain three questions that need empirical exploration. First, Wright's theory and our studies indicated that the average additive effect on fitness of single genes will vary if measured in the different genetic backgrounds of different local demes (Wade 2001, 2002). We need to measure and quantify the variation in the fitness effects of single genes on different genetic backgrounds (Figure 10.9) as is now being done in some labs. For example, after comparing the effects of genes in *D. melanogaster* across several different genetic backgrounds, Huang et al. (2012, p. 15,553) recently concluded that epistasis is "a principal factor that determines variation for quantitative traits." They further emphasized that it is important to understand epistasis in order to enhance "our understanding of the genetic basis of evolutionarily and clinically im-

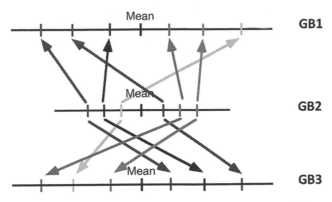

FIGURE 10.9 This is a schematic illustration of the change in the effects of different genes relative to the mean phenotype (indicated by hash marks on the chromosomes) in different genetic backgrounds. This figure has been modified from C. J. Goodnight (2014) (https://blog .uvm.edu/cgoodnig/).

portant traits." In a similar vein, Dr. T. MacKay (2014, p. 22) reviewed studies of gene interactions in model organisms and concluded that "epistasis is common, and that additivity can be an emergent property of underlying genetic interaction networks."

We also need to know how different one genetic background must be from another before the variability in a gene's effect on fitness is sufficiently large that the effect changes sign, increasing fitness in one genetic background but decreasing it in another. If the effect of a single gene on fitness changes sign as a function of variation in genetic background, we need to know the range of F_{ST} values where such changes begin to occur. We still know too little about metapopulations to connect the variation in background, which we describe by F_{ST}, with the effects of single genes. Could *any* gene be a speciation gene under the right conditions of genetic background or is there a special class of genes particularly sensitive to variation in genetic background?

If classical gene interactions are as common as Wright asserted, then the additive genetic variance for fitness should increase when populations at a genetic equilibrium are inbred or subdivided into metapopulations. We have evidence from the increase in statistical genetic parameters, such as h^2 and g^2, that this is so in our experimental populations (Wade 1985a; Wade et al. 1996; Wade and Griesemer 1998; Drury and Wade 2011), but we do not yet have evidence at the level of the single genes that underlie these statistical genetic parameters. Lastly, we should expect that the genes responsible for a response to individual selection in small populations might vary from deme to deme across a metapopulation, even for uniformly imposed selection.

When we turn to the investigation of interactions between individuals, there are many more avenues for both theoretical and experimental investigation. In our theoretical work, we have considered the theory of maternal genetic effects as a guide to the study of indirect genetic effects in general (Wade 1998; Wolf and Wade 2009). Maternal genetic effects are genes in the maternal genome that affect offspring phenotypes. Effects of genes in maternal genomes appear as part of the additive genetic variance among their offspring. Their evolutionary dynamics, even without interactions, are very different from those of genes with direct effects. The relative evolutionary influences of genes in the offspring genome and those in the maternal genome change with inbreeding and with population genetic subdivision. Moreover, inbreeding amplifies the importance of maternal genetic effects in predicting the trajectory and outcome of evolution. Fortunately, we have found that selection leaves a distinctive signature on the DNA sequence variation of genes with only maternal effects (Demuth and Wade 2007; Wade, Priest, and Cruickshank 2009) and our theoretical investigations into social genetics predict similar patterns (Van Dyken, Linksvayer, and Wade 2011; Van Dyken and Wade 2012), which are not yet confirmed by empirical data.

In short, much more research is needed to support and extend the evolutionary processes discussed in the preceding pages.

Acknowledgments

Research is a compelling type of teamwork and, throughout this monograph, I have tried to document the tremendous input of creativity, industry, and community that my graduate students and postdoctoral fellows contributed to research in the Wade lab. My great debt to them as scientists and friends cannot be overstated or repaid. Vetting ideas with them made my choice of experiments better and their results more robust.

I am greatly indebted to my wife, Debra L. Rush-Wade, and children, Catherine, Megan, and Travis Wade, for their generous support and enduring patience throughout the writing of this book.

Chris Caruso, Hafiz Maherali, and Dena Smith, fellow sabbatical scholars and soiree planners, helped keep me on track and off the potentially debilitating "sabbatical spiral" during my long stay at NESCent. I am especially grateful to Chris, whom I imposed on almost daily to read chapter drafts, some polished and some ill-conceived. Her support for my project in its earliest days propelled me to accomplish much more on sabbatical than I would have otherwise.

Several others read and commented on the manuscript in progress, including Steve Shuster, Rasmus Winther, David McCauley, Curt Lively, Dave van Dyken, Jeff Demuth, Jake Moorad, Yaniv Brandvain, Devin Drown, Kevin Preuss, and

Doug Drury. In particular, Rasmus' early comments helped shape the overall structure of the book. Jim Zehr read the first couple chapters and encouraged me to continue with the poker metaphor as a means of making interactions and their effects clearer to nontechnical readers. In the summer and fall of 2014, Richard Prum, the William Robertson Coe Professor of Ornithology and Head Curator of Vertebrate Zoology at the Peabody Museum of Natural History at Yale University, startled me with his enthusiastic response to the first two chapters. His unexpected support pushed me to work more diligently toward the book's completion. In a similar vein, I am especially indebted to my colleague and friend Lisa A. Lloyd, the Arnold and Maxine Tanis Chair of History and Philosophy of Science. Lisa critically read drafts of my first several chapters in the fall of 2014, commented insightfully on them in great detail, and saw value in the project at a time when my enthusiasm for it was flagging. Without her collegial support, continued interest, avid discussion, and insightful reading of every chapter, the manuscript would be much diminished.

Lastly, I would like to acknowledge Christie Henry, Editorial Director for Sciences at the University of Chicago Press, for her enthusiastic response to my inquiry about the project. Mary Corrado, a senior manuscript editor for the Press, also greatly improved the manuscript with her perceptive copyediting.

This work was supported by an award from the National Science Foundation Opportunities for Promoting Understanding through Synthesis program, by a Senior Sabbatical Fellowship from the National Evolutionary Synthesis Center, and by Indiana University Bloomington through its Sabbatical Leave program. I am grateful for the long history of research support from both the National Institutes of Health and the National Science Foundation. These institutions not only allowed me to pursue the work synthesized here but also supported the developing careers of the graduate students and postdoctoral fellows mentioned throughout the book.

Reference List

Agrawal, A. F., E. D. Brodie III, and M. J. Wade. 2001. On indirect genetic effects in structured populations. American Naturalist 158:308–323.

Arnold, S. J., and T. Halliday. 1992. Multiple mating by females: the design and interpretation of selection experiments. Animal Behaviour 43:178–179.

Arnold, S. J., and M. J. Wade. 1984. On the measurement of selection in natural and laboratory populations: theory. Evolution 38:709-719.

Barash, D. P. 1977. Sociobiology and behavior. Elsevier, New York.

Barash, D. P. 1979. The whisperings within: evolution and the origin of human nature. Harper and Row, New York.

Barton, N. H. 1992. On the spread of new gene combinations in the third phase of Wright's shifting-balance. Evolution 46:551–557.

Barton, N. H. 1995. Linkage and the limits to natural selection. Genetics 140:821–841.

Barton, N. H., and S. Rouhani. 1991. The probability of fixation of a new karyotype in a continuous population. Evolution 54:499–517.

Barton, N. H., and S. Rouhani. 1993. Adaptation and the "shifting balance." Genetical Research 61:57–74.

Bateman, A. J. 1948. Intra-sexual selection in *Drosophila*. Heredity 2:349–368.

Bell, G. 1978. Group selection in structured populations. American Naturalist 112:389–399.

Biscarini, F., H. Bovenhuis, J. Van Der Poel, T. B. Rodenburg, A. P. Jungerius, and J. A. M. Van Arendonk. 2010. Across-line SNP association study for direct and associative effects on feather damage in laying hens. Behavior Genetics 40:715–727.

Bittles, A. H., and M. L. Black. 2009. Consanguinity, human evolution, and complex diseases. Proceedings of the National Academy of Sciences USA 107 (suppl 1): 1779–1786.

Boake, C. R., ed. 1994. Quantitative genetic studies of behavioral evolution. University of Chicago Press, Chicago.

Boake, C. R., and R. R. Capranica. 1982. Aggressive signal in "courtship" chirps of a gregarious cricket. Science 218:580–582.

Brandon, R. 1982. The levels of selection. In: PSA: Proceedings of the biennial meeting of the Philosophy of Science Association, 315–323. University of Chicago Press, Chicago.

Brandvain, Y., C. J. Goodnight, and M. J. Wade. 2011. Disequilibria between organelle and symbiont genomes with random genetic drift. Genetics 189:397–404.

Breden, F. J., and M. J. Wade. 1981. Inbreeding and evolution by kin selection. Ethology and Sociobiology 2:3–16.

Breden, F. J., and M. J. Wade. 1985. The effects of group size and cannibalism rate on larval growth and survivorship in *Plagiodera versicolora*. Entomography 3:455–463.

Breden, F. J., and M. J. Wade. 1987. An experimental study of the effect of group size on larval growth and survivorship in the imported willow leaf beetle, *Plagiodera versicolora* (Coleoptera: Chrysomelidae). Environmental Entomology 16:1082–1086.

Breden, F. J., and M. J. Wade. 1989. Selection within and between kin groups in the imported willow leaf beetle, *Plagiodera versicolora*. American Naturalist 134:35–50.

Broyles, S. B., and R. Wyatt. 1990. Paternity analysis in a natural population of *Asclepias exaltata*: multiple paternity, functional gender, and the "pollen-donation hypothesis." Evolution 44:1454–1468.

Buri, P. 1956. Gene frequency in small populations of mutant *Drosophila*. Evolution 10:367–402.

Carson, H. L. 1961. Relative fitness of genetically open and closed experimental populations of *Drosophila robusta*. Genetics 46:553–567.

Cassidy, J. 1978. Philosophical aspects of the group selection controversy. Philosophy of Science 45:575–594.

Chapman, M., and G. Hausfater. 1979. The reproductive consequences of infanticide in langurs: a mathematical model. Behavioral Ecology and Sociobiology 5:227–240.

Charlesworth, B. 1970. Selection in populations with overlapping generations, I: the use of Malthusian parameters in population genetics. Theoretical Population Biology 1:352–370.

Charlesworth, B. 1972. Selection in populations with overlapping generations, III: conditions for genetic equilibrium. Theoretical Population Biology 3:377–395.

Charlesworth, B., and D. Charlesworth. 2010. Elements of evolutionary genetics. Roberts, Greenwood Village, CO.

Cheng, H. W., and R. L. Dennis. 2011. The dopaminergic system and aggression in laying hens. Poultry Science 90:2440–2448.

Cole, L. 1954. The population consequences of life history phenomena. Quarterly Review of Biology 29:103–137.

Colwell, R. K. 1981. Group selection is implicated in the evolution of female-biased sex ratios. Nature 290:401–404.

Costa, J. T., III, and K. G. Ross. 1993. Seasonal decline in intracolony genetic relatedness in eastern tent caterpillars: implications for social evolution. Behavioral Ecology and Sociobiology 32:47–54.

Coyne, J. A., N. H. Barton, and M. Turelli. 1997. Perspective: a critique of Sewall Wright's shifting balance theory of evolution. Evolution 51:643–671.

Coyne, J. A., N. H. Barton, and M. Turelli. 2000. Is Wright's shifting balance process important in evolution? Evolution 54:306–317.

Craig, D. M. 1982. Group selection versus individual selection: an experimental analysis. Evolution 36:271–282.

Craig, J. V., and W. M. Muir. 1996. Group selection for adaptation to multiple-hen cages: beak-related mortality, feathering, and body weight responses. Poultry Science 75:294–302.

Crow, J. F. 1958. Some possibilities for measuring selection intensities in man. Human Biology 30:1–13.

Crow, J. F. 1962. Population genetics: selection. In: Methodology in human genetics, edited by W. J. Burdette, 53–75. Holden-Day, San Francisco.

Crow, J. F. 1991. Was Wright right? Science 253:973.

Crow, J. F. 2010. On epistasis: why it is unimportant in polygenic directional selection. Philosophical Transactions of the Royal Society B 365:1241–1244.

Crow, J. F., and K. Aoki. 1982. Group selection for a polygenic behavioral trait: a differential proliferation model. Proceedings of the National Academy of Sciences USA 79:2628–2631.

Crow, J. F., and M. Kimura.1970. An introduction to population genetics theory. Harper and Row, New York.

Crow, J. F., and N. E. Morton. 1955. Measurement of gene frequency drift in small populations. Evolution 9:202–214.

Cruickshank, T., and M. J. Wade. 2008. Microevolutionary support for a developmental hourglass: gene expression patterns shape sequence variation and divergence in *Drosophila*. Evolution and Development 10:583–590.

Darwin, C. 1859. On the origin of species. John Murray, London.

Darwin, C. 1871. The descent of man and selection in relation to sex. John Murray, London.

Dawkins, R. 1976. The selfish gene. Oxford University Press, New York.

Dawkins, R. 1978. Replicator selection and extended phenotype. Zeitschrift fur Tierpsychologie 47:61–76.

Dawkins, R. 1979. Twelve misunderstandings of kin selection. Zeitschrift für Tierpsychologie 51:184–200.

Dawkins, R. 1982. The extended phenotype. Oxford University Press, New York.

Dawkins, R., and J. R. Krebs. 1978. Animal signals: information or manipulation. Behavioural Ecology 3:282–309.

Dawson, P. S. 1967. Developmental rate and competitive ability in *Tribolium*, III: competition in unfavorable environments. Journal of Stored Products Research 3:193–198.

Dawson, P. S. 1968. Xenocide, suicide, and cannibalism in flour beetles. American Naturalist 102:97–105.

Demuth, J., and M. J. Wade. 2007. Maternal expression increases the rate of *bicoid* evolution by relaxing selective constraint. Genetica 129:37–43.

Dobzhansky, T. 1937. Genetics and the origin of species. Columbia University Press, New York.

Dobzhansky, T., and O. Pavlovsky. 1957. An experimental study of interaction between genetic drift and natural selection. Evolution 11:311–319.

Drown, D. M., and M. J. Wade. 2014. Runaway coevolution: adaptation to heritable and non-heritable environments. Evolution 68:3039–3046.

Drown, D. M., P. C. Zee, Y. Brandvain, and M. J. Wade. 2013. Evolution of transmission mode in obligate symbionts. Ecological Evolutionary Research 15:43–59.

Drury, D. W., and M. J. Wade. 2011. Genetic variation in fitness between sympatric and allopatric backgrounds in the red flour beetle, *Tribolium castaneum*. Journal of Evolutionary Biology 24:168–176.

Eberhard, W. G. 1996. Female control: sexual selection by cryptic female choice. Princeton University Press, Princeton, NJ.

Ellen, E. D., J. Visscher, J. A. Van Arendonk, and P. Bjima. 2008. Survival of laying hens:

genetic parameters for direct and associative effects in three purebred layer lines. Poultry Science 87:233–239.

Estoup, A., M. Solignac, and J. M. Cornuet. 1994. Precise assessment of the number of patrilines and of genetic relatedness in honeybee colonies. Proceedings of the Royal Society of London B 258:1–7.

Ewens, W. J. 2000. The mathematical foundations of population genetics. In: Evolutionary genetics, edited by R. S. Singh and C. B. Krimbas, 24–40. Cambridge University Press, Cambridge, UK.

Falconer, D. S. 1965. The inheritance of liability to certain diseases, estimated from the incidence among relatives. Annals of Human Genetics 29:51–76.

Falconer, D. S. 1989. Introduction to quantitative genetics. 3rd ed. Longmans Green, Harlow, UK.

Falconer, D. S., and T. F. C. Mackay. 1996. Introduction to quantitative genetics. 4th ed. Longman, Essex, UK.

Feldman, M. W., F. B. Christiansen, and U. Liberman. 1983. On some models of fertility selection. Genetics 105:1003–1010.

Feller, W. 1971. An introduction to probability theory and its applications. Vol. 1. Wiley and Sons, New York.

Feller, W. 1971. An introduction to probability theory and its applications. Vol. 2. Wiley and Sons, New York.

Fisher, R. A. 1918. The correlation between relatives on the supposition of Mendelian inheritance. Transactions of the Royal Society of Edinburgh 52:399–433.

Fisher, R. A. 1930. The genetical theory of natural selection. Clarendon Press, Oxford, UK.

Fisher, R. A. 1958. The genetical theory of natural selection. Dover, New York.

Fisher, R. A., and E. B. Ford. 1950. The "Sewall Wright" effect. *Heredity* 4:117–119.

Frankham, R. 1995. Conservation genetics. Annual Review of Genetics 29:305–327.

Franklin, I., and R. C. Lewontin. 1970. Is the gene the unit of selection? Genetics 65:707–734.

Futuyma, D. J. 1970. Variation in genetic response to interspecific competition in laboratory populations of *Drosophila*. American Naturalist 104:239–252.

Ghiselin, M. T. 1974. Economy of nature and the evolution of sex. University of California Press, Berkeley.

Ghoul, M., A. S. Griffin, and S. A. West. 2014. Toward an evolutionary definition of cheating. Evolution 68:318–331.

Gillespie, J. H. 1998. Population genetics: a concise guide. Johns Hopkins University Press, Baltimore, MD.

Goodnight, C. J. 1985. The influence of environmental variation on group and individual selection in a cress. Evolution 39:545–558.

Goodnight, C. J. 1987. On the effect of founder events on epistatic genetic variance. Evolution 41:80–91.

Goodnight, C. J. 1988. Epistasis and the effect of founder events on the additive genetic variance. Evolution 42:441–454.

Goodnight, C. J. 1990a. Experimental studies of community evolution, I: the response to selection at the community level. Evolution 44:1614–1624.

Goodnight, C. J. 1990b. Experimental studies of community evolution, II: the ecological basis of the response to community selection. Evolution 44:1625–1636.

Goodnight, C. J. 1995. Epistasis and the increase in additive genetic variance: implications for phase 1 of Wright's shifting-balance process. Evolution 49:502–511.

Goodnight, C. J., and M. J. Wade. 2000. The ongoing synthesis: a reply to Coyne et al. (1999). Evolution 54:317–324.

Grant, V. 1978. Kin selection: a critique. Biologisches Zentralblatt 97:385–392.

Griffin, A. S., and S. A. West. 2002. Kin selection: fact and fiction. Trends in Ecology and Evolution 17:16–21.

Griffing, B. 1967. Selection in reference to biological groups, I: individual and group selection applied to populations of unordered groups. Australian Journal of Biological Science 20:127–139.

Haldane, J. B. S. 1932. The causes of evolution. Cornell University Press, Itaca, NY.

Hamilton, W. D. 1963. The evolution of altruistic behavior. American Naturalist 97:354–356.

Hamilton, W. D. 1964a. The genetical evolution of social behaviour, I. Journal of Theoretical Biology 7:1–16.

Hamilton, W. D. 1964b. The genetical evolution of social behaviour, II. Journal of Theoretical Biology 7:17–52.

Hamilton, W. D. 1970. Selfish and spiteful behaviour in an evolutionary model. Nature 2228:1218–1220.

Hamilton, W. D. 1971. Geometry of the selfish herd. Journal of Theoretical Biology 31:295–311.

Hamilton, W. D. 1975. Innate social aptitudes of man: an approach from evolutionary genetics. In: Biosocial anthropology, edited by R. Fox, 133–153. Malaby Press, London, UK.

Hartl, D. 1980. Principles of population genetics. Sinauer Associates, Sunderland, MA.

Hedrick, P. W., M. E. Ginevana, and E. P. Ewing. 1976. Genetic polymorphism in heterogeneous environments. Annual Review of Ecology and Systematics 7:1–32.

Hedrick, P. W., and S. T. Kalinowski. 2000. Inbreeding depression and conservation biology. Annual Review of Ecology and Systematics 31:139–162.

Hill, W. G., and A. Caballero. 1992. Artificial selection experiments. Annual Review of Ecology and Systematics 23:287–310.

Hill, W. G., and A. Robertson. 1966. The effect of linkage on limits to artificial selection. Genetical Research 8:269–294.

Hill, W. G., M. E. Goddard, and P. M. Visscher. 2008. Data and theory point to mainly additive genetic variance for complex traits. PLOS Genetics 4:e1000008.

Holland, J. H. 1992. Genetic algorithms. Scientific American 267:66–72.

Huang, W., S. Richards, M. A. Carbone, et al. 2012. Epistasis dominates the genetic architecture of Drosophila quantitative traits. Proceedings of the National Academy of Sciences, USA 109:15553–15559.

Huffaker, C. B. 1958. Experimental studies on predation: dispersion factors and predator-prey oscillations. Hilgardia 27:343–384.

Huffaker, C. B., S. G. Herman, and K. P. Shea. 1963. Experimental studies on predation: complex dispersion and levels of food in an acarine predator-prey interaction. Hilgardia 34:305–330.

Hull, D. L. 1980. Individuality and selection. Annual Review of Ecology and Systematics 11:311–332.

Hull, D. L. 1988. Interactors versus vehicles. In: The role of behavior in evolution, edited by H. C. Plotkin, 19–50. MIT Press, Cambridge, MA.

Jacobs, M., and M. J. Wade. 2003. A synthetic review of the theory of gynodioecy. American Naturalist 161:837–851.

Jaenike, J. 2001. Sex chromosome meiotic drive. Annual Review of Ecology and Systematics 32:25–49.

Janzen, D. 1968. Host plants as islands in evolutionary and contemporary time. American Naturalist 102:592–595.

Janzen, D. H. 1973. Host plants as islands, II: competition in evolutionary and contemporary time. American Naturalist 107:786–789.

Johnson, N. 2008. Sewall Wright and the development of shifting balance theory. Nature Education 1:52.

Jolivet, P. 2008. Cycloalexy. In: Encyclopedia of entomology, 2nd ed., edited by J. L. Capinera, 1139–1140. Springer, Dordrecht, The Netherlands.

Keller, L., ed. 1999. Levels of selection in evolution. Princeton University Press, Princeton.

Kerr, W. E., and S. Wright. 1954. Experimental studies of the distribution of gene frequencies in very small populations of *Drosophila melanogaster*, I: forked. Evolution 8:172–177.

Kimura, M. 1968. Evolutionary rate at the molecular level. Nature 217:624–625.

Kimura, M. 1985. The neutral theory of molecular evolution. Cambridge University Press, New York.

King, J. L., and T. H. Jukes. 1969. Non-Darwinian evolution. Science 164:788–798.

Lack, D. 1947. The significance of clutch-size. Ibis 89:302–352.

Lack, D. 1948. The significance of clutch-size, III: some interspecific comparisons. Ibis 90:25–45.

Lande, R. 1976. Natural selection and random genetic drift in phenotypic evolution. Evolution 30:314–334.

Lande, R. 1977. Statistical tests for natural selection on quantitative characters. Evolution 31:442–444.

Lande, R., and S. J. Arnold. 1983. The measurement of selection on correlated characters. Evolution 37:1210–1226.

Larson, A., D. B. Wake, and K. P. Yanev. 1984. Measuring gene flow among populations having high levels of genetic fragmentation. Genetics 106:293–308.

Lerner, I. M. 1958. The genetic basis of selection. Wiley and Sons, New York.

Levin, B. R., and W. L. Kilmer. 1974. Interdemic selection and the evolution of altruism: a computer simulation study. Evolution 28:527–545.

Levins, R. 1962. Theory of fitness in a heterogeneous environment, I: the fitness set and adaptive function. American Naturalist 96:361–373.

Levins, R. 1967. Evolution in changing environments. Princeton University Press, Princeton.

Levins, R. 1970. Extinction. Lectures on mathematics in the life sciences. 2:75–107.

Lewontin, R. C. 1955. The effects of population density and composition on viability in *Drosophila melanogaster*. Evolution 9:27–41.

Lewontin, R. C. 1965. Selection for colonizing ability. In: The genetics of colonizing species: proceedings of the First International Union of Biological Sciences Symposia on General Biology, 77–94. Academic Press, New York.

Lewontin, R. C. 1970a. On the irrelevance of genes. In: Towards a theoretical biology, edited by C. H. Waddington. Vol. 3: Drafts, 63–72. Edinburgh University Press, Edinburgh, UK.

Lewontin, R. C. 1970b. The units of selection. Annual Review of Ecology and Systematics 1:1–18.

Lewontin, R. C. 1974. The genetic basis of evolutionary change. Columbia University Press, New York.

Lewontin, R. C., and J. L. Hubby. 1966. A molecular approach to the study of genic heterozygosity in natural populations, II: amount of variation and degree of heterozygosity in natural populations of *Drosophila pseudoobscura*. Genetics 54:595–609.

Lewontin, R. C., and K. I. Kojima. 1960. The evolutionary dynamics of complex polymorphisms. Evolution 14:458–472.

Lewontin, R. C., and Y. Matsuo. 1963. Interaction of genotypes determining viability in *Drosophila busckii*. Proceedings of the National Academy of Sciences USA 49:270–278.

Linksvayer, T., and M. J. Wade. 2009. Genes with social effects are expected to harbor more sequence variation within and between species. Evolution 63:1685–1696.

Lloyd, E. A. 1984. A semantic approach to the structure of population genetics. Philosophy of Science 51:242–264.

Lloyd, E. A. 1988. The structure and confirmation of evolutionary theory. Princeton University Press, Princeton.

Lloyd, E. A., and M. W. Feldman. 2002. Commentary: evolutionary psychology: a view from evolutionary biology. Psychological Inquiry 13:150–156.

Lloyd, E. A., R. C. Lewontin, and M. W. Feldman. 2008. The generational cycle of state spaces and adequate genetical representation. Philosophy of Science 75:140–156.

Lloyd, M. 1968. Self regulation of adult numbers by cannibalism in two laboratory strains of flour beetles (*Tribolium castaneum*). Ecology 49:245–259.

Loewe, L., and W. G. Hill. 2010. The population genetics of mutations: good, bad and indifferent. Philosophical Transactions of the Royal Society B 365:1153–1167.

López-Fanjul, C. 1989. Tests of theory by selection experiments. In: Evolution and animal breeding: reviews on molecular and quantitative approaches in honour of Alan Robertson, edited by W. G. Hill and T. F. C. Mackay, 129–133. C. A. B. International, Wallingford, UK.

Luckinbill, L. S. 1978. r and K selection in experimental populations of *Escherichia coli*. Science 202:1201–1203.

MacArthur, R. H. 1972. Geographical ecology: patterns in the distribution of species. Harper and Rowe, New York.

MacArthur, R. H., and E. O. Wilson. 1967. The theory of island biogeography. Princeton University Press, Princeton, NJ.

Mackay, T. F. 2014. Epistasis and quantitative traits: using model organisms to study gene-gene interactions. Nature Reviews Genetics 15:22–33.

Mallet, J., and M. Joron. 1999. Evolution of diversity in warning color and mimicry: polymorphisms, shifting balance, and speciation. Annual Review of Ecology and Systematics 30:201–233.

Maruyama, T. 1970. Effective number of alleles in a subdivided population. Theoretical Population Biology 1:273–306.

Mather, K., and B. J. Harrison. 1949. The manifold effect of selection. Heredity 3:131–162.

Maynard Smith, J. 1964. Group selection and kin selection. Nature 201:1145–1147.

Maynard Smith J. 1976. Group selection. Quarterly Review of Biology 61: 277–283.

Maynard Smith, J. 1978. The evolution of sex. Cambridge University Press, Cambridge, UK.

Maynard Smith, J. 1982. Evolution and the theory of games. Cambridge University Press, Cambridge, UK.

Maynard Smith, J., and M. Slatkin. 1973. The stability of predator-prey systems. Ecology 54:384–391.

Mayr, E. 1942. Systematics and the origin of species, from the viewpoint of a zoologist. Harvard University Press, Cambridge, MA.

Mayr, E. 1988. Toward a new philosophy of biology: observations of an evolutionist. Harvard University Press, Cambridge, MA.

McCauley, D. E. 1978. Demographic and genetic responses of two strains of *Tribolium castaneum* to a novel environment. Evolution 32:398–415.

McCauley, D. E. 1979. Geographic variation in body size and its relation to the mating structure of Tetraopes populations. Heredity 42:143–148.

McCauley, D. E., and R. O'Donnell. 1984. The effect of multiple mating on genetic relatedness in larval aggregations of the imported willow leaf beetle (*Plagiodera versicolora*, Coleoptera: Chrysomelidae). Behavioral Ecology and Sociobiology 15:287–291.

McCauley, D. E., and D. R. Taylor. 1997. Local population structure and sex ratio: evolution in gynodioecious plants. American Naturalist 150:406–419.

McCauley, D. E., and M. J. Wade. 1978. Female choice and the mating structure of a natural population of the soldier beetle, *Chauliognathus pennsylvanicus*. Evolution 32:771–775.

McCauley, D. E., and M. J. Wade. 1980. Group selection: the genetic and demographic basis for the phenotypic differentiation of small populations of *Tribolium castaneum*. Evolution 34:813–821.

McCauley, D. E., and M. J. Wade. 1981. The populational effects of inbreeding in *Tribolium*. Heredity 46:59–67.

McCauley, D. E., M. J. Wade, F. J. Breden, and M. Wohltman. 1988. Kin selection: geographic variation in relatedness in the imported willow leaf beetle, *Plagiodera versicolora*. Evolution 42:184–192.

Mertz, D. B. 1969. Age-distribution and abundance in populations of flour beetles, I: experimental studies. Ecological Monographs 39:1–31.

Mertz, D. B., and D. E. McCauley. 1980. The domain of laboratory ecology. Synthese 43:95–110.

Michod, R. E. 1980. Evolution of interactions in family-structured populations: mixed mating models. Genetics 96:275–296.

Michod, R. E. 1982. The theory of kin selection. Annual Review of Ecology and Systematics 13:23–55.

Moore, A. J., J. B. Wolf, and E. D. Brodie III. 1998. The influence of direct and indirect genetic effects on the evolution of behavior: sexual and social selection meet maternal effects. In: Maternal effects as adaptations, edited by T. A. Mousseau and C. Fox, 22–41. Oxford University Press, Oxford, UK.

Mueller, L. D., and F. J. Ayala. 1981. Trade-off between r-selection and K-selection in *Drosophila* populations. Proceedings of the National Academy of Sciences USA 78:1303–1305.

Muir, W. M. 1996. Group selection for adaptation to multiple-hen cages: selection program and direct responses. Poultry Science 75:447–458.

Muller, H. J. 1950. Our load of mutations. American Journal of Human Genetics 2:111–176.

Mumme, R. L., W. D. Koenig, and F. A. Pitelka. 1983. Mate guarding in the acorn woodpecker: within-group reproductive competition in a cooperative breeder. Animal Behaviour 31:1094–1106.

Nathanson, M. R. 1975. The effect of resource limitation on competing populations of flour beetles, *Tribolium spp.* Bulletin of Entomological Research 65:1–12.

Nei, M., and M. W. Feldman. 1972. Identity of genes by descent within and between populations under mutation and migration pressures. Theoretical Population Biology 3:460–465.

Neyman, J., T. Park, and E. L. Scott. 1956. Struggle for existence: the *Tribolium* model: biological and statistical aspects. In: Proceedings of the Third Berkeley Symposium on Mathematical Statistics and Probability 4:41–79. University of California Press, Berkeley.

Nonacs, P., and R. Hager. 2011. The past, present and future of reproductive skew theory and experiments. Biological Reviews 86:271–298.

O'Farrell, P. H. 1975. High resolution two-dimensional electrophoresis of proteins. Journal of Biological Chemistry 250:4007–4021.

Park, T. 1932. Studies in population physiology: the relation of numbers to initial population growth in the flour beetle *Tribolium confusum* Duval. Ecology 13:172–181.

Park, T. 1948. Experimental studies of interspecies competition, I: competition between populations of the flour beetles, *Tribolium confusum* Duval and *Tribolium castaneum* Herbst. Ecological Monographs 18:265–307.

Park, T. 1954. Experimental studies of interspecies competition, II: temperature, humidity, and competition in two species of *Tribolium*. Physiological Zoology 27:177–238.

Park, T., P. H. Leslie, and D. B. Mertz. 1964. Genetic strains and competition in populations of *Tribolium*. Physiological Zoology 37:97–162.

Park, T., J. R. Ziegler, D. L. Ziegler, and D. B. Mertz. 1974. The cannibalism of eggs by *Tribolium* larvae. Physiological Zoology 47:37–58.

Pianka, E. R. 1970. On r-and K-selection. American Naturalist 104:592–597.

Pianka, E. R. 1978. Evolutionary ecology. Oxford University Press, Oxford, UK.

Pianka, E. R. 2011. Evolutionary ecology. E. R. Pianka, e-book.

Polis, G. A. 1981. The evolution and dynamics of intraspecific predation. Annual Review of Ecology and Systematics 12:225–251.

Polis, G. A. 1984. Intraspecific predation and "infant killing" among invertebrates. In: Infanticide: comparative and evolutionary perspectives, edited by G. Hausfater and S. Blaffer Hrdy, 87–104. Aldine, New York.

Price, G. R. 1972. Extension of covariance selection mathematics. Annals of Human Genetics 35:485–490.

Queller, D. C., J. E. Strassmann, and C. R. Hughes. 1993. Microsatellites and kinship. Trends in Ecology and Evolution 8:285–288.

Reddingius, J. 1971. Gambling for existence: a discussion of some theoretical problems in animal population ecology. Acta Biological Theory 20:S1–208.

Rice, W. R. 2013. Nothing in genetics makes sense except in light of genomic conflict. Annual Review of Ecology, Evolution, and Systematics 44:217–237.

Ricklefs, R. E. 1975. The evolution of cooperative breeding in birds. Ibis 117:531–534.

Rittle-Johnson, B., and J. R. Star. 2007. Does comparing solution methods facilitate conceptual and procedural knowledge? An experimental study on learning to solve equations. Journal of Educational Psychology 99:561–574.

Robertson, A. 1966. A mathematical model of the culling process in dairy cattle. Animal Production 8:95–108.

Robertson, A. 1967. Animal breeding. Annual Review of Genetics 1:295–312.

Robertson, F. W., and E. C. R. Reeve. 1952. Heterozygosity, environmental variation and heterosis. Nature 170:286.

Rosvall, O., D. Lindgren, and T. J. Mullin. 1998. Sustainability robustness and efficiency of a multi-generation breeding strategy based on within-family clonal selection. Silvae Genetica 47:307–320.

Roughgarden, J. 1971. Density-dependent natural selection. Ecology 52:453–468.

Ruse, M. 1980. Charles Darwin and group selection. Annals of Science 37:615–630.

Santos, M., and E. Szathmáry. 2008. Genetic hitchhiking can promote the initial spread of strong altruism. BMC Evolutionary Biology 8:281.

Shaw, R. G. 1986. Response to density in a wild population of the perennial herb *Salvia lyrata*: variation among families. Evolution 40:492–505.

Sherman, P. W. 1981. Kinship, demography, and Belding's ground squirrel nepotism. Behavioral Ecology and Sociobiology 8:251–259.

Shuster, S. M., and M. J. Wade. 2003. Mating systems and mating strategies. Princeton University Press, Princeton.

Simberloff, D., and E. O. Wilson. 1970. Experimental zoogeography of islands: a two-year record of colonization. Ecology 51:934–937.

Simpson, G. G. 1944. Tempo and mode in evolution. Columbia University Press, New York.

Skipper, R. A. 2002. The persistence of the R. A. Fisher–Sewall Wright controversy. Biology and Philosophy 17:341–367.

Slatkin, M. 1972. On treating the chromosome as the unit of selection. Genetics 72:157–168.

Slatkin, M. 1974. Competition and regional coexistence. Ecology 55:128–134.

Slatkin, M. 1981. Fixation probabilities and fixation times in a subdivided population. Evolution 35:477–488.

Slatkin, M. 1985. Gene flow in natural populations. Annual Review of Ecology and Systematics 16:393–430.

Sober, E. 1984. The nature of selection. MIT Press, Cambridge, MA.

Sober, E., and R. C. Lewontin. 1982. Artifact, causes, and genic selection. Philosophy of Science 49:147–176.

Stevens, L., and D. E. McCauley. 1989. Mating prior to overwintering in the imported willow leaf beetle, *Plagiodera versicolora* (Coleoptera: Chrysomelidae). Ecological Entomology 14:219–223.

Stevens, L., and M. J. Wade. 1990. Cytoplasmically inherited reproductive incompatibility in *Tribolium* flour beetles: the rate of spread and effect on population size. Genetics 124:367-372.

Sved, J. A., and B. D. H. Latter. 1977. Migration and mutations in stochastic models of gene frequency change, I: the island model. Journal of Mathematical Biology 5:61–73.

Taylor, C. E., and C. Condra. 1980. r- and K-selection in *Drosophila pseudoobscura*. Evolution 34:1183–1193.

Toquenaga, Y., and M. J. Wade. 1996. Sewall Wright meets artificial life: the origin and maintenance of evolutionary novelty. Trends in Ecology and Evolution 11:478–482.

Trivers, R. L., and H. Hare. 1976. Haplodiploidy and the evolution of the social insect. Science 191:249–263.

Uyenoyama, M. K. 1979. Evolution of altruism under group selection in large and small populations in fluctuating environments. Theoretical Population Biology 15:58–85.

Van Dyken, J. D., and M. J. Wade. 2010. Quantifying the evolutionary consequences of conditional gene expression in time and space. Genetics 184:557–570.

Van Dyken, J. D., and M. J. Wade. 2012. Detecting the molecular signature of social conflict: theory and a test with bacterial quorum sensing genes. American Naturalist 179:436–450.

Van Dyken, J. D., T. Linksvayer, and M. J. Wade. 2011. Kin selection–mutation balance: a model for the origin, maintenance, and consequences of social cheating. American Naturalist 177:288–300.

Via, S. 1984a. The quantitative genetics of polyphagy in an insect herbivore, I: genotype-environment interaction in larval performance on different host plant species. Evolution 38:881–895.

Via, S. 1984b. The quantitative genetics of polyphagy in an insect herbivore, II: genetic correlations in larval performance within and among host plants. Evolution 38:896–905.

Wade, M. J. 1976. Group selection among laboratory populations of *Tribolium*. Proceedings of the National Academy of Science USA 73:4604 4607.

Wade, M. J. 1977. An experimental study of group selection. Evolution 31:134–153.

Wade, M. J. 1978a. A critical review of the models of group selection. Quarterly Review of Biology 53:101–114.

Wade, M. J. 1978b. Kin selection: a classical approach and a general solution. Proceedings of the National Academy of Science USA 75:6154–6158.

Wade, M. J. 1978c. The selfish gene: a review. Evolution 32:220–221.

Wade, M. J. 1979a. The primary characteristics of *Tribolium* populations group selected for increased and decreased population size. Evolution 33:749–764.

Wade, M. J. 1979b. Sexual selection and variance in reproductive success. American Naturalist 114:742–747.

Wade, M. J. 1979c. The evolution of social interactions by family selection. American Naturalist 113:399–417.

Wade, M. J. 1980a. An experimental study of kin selection. Evolution 34:844–855.

Wade, M. J. 1980b. Effective population size: the effects of sex, genotype and density of the mean and variance of offspring numbers in the flour beetle, *Tribolium castaneum*. Genetical Research 36:1–10.

Wade, M. J. 1980c. Group selection, population growth rate, and competitive ability in the flour beetle, *Tribolium* spp. Ecology 61:1056–1064.

Wade, M. J. 1980d. Kin selection: its components. Science 210:665–667.

Wade, M. J. 1980e. Wright's view of evolution. Science 207:173–174.

Wade, M. J. 1982a. The effect of multiple inseminations on the evolution of social behavior in diploid and haplodiploid organisms. Journal of Theoretical Biology 95:351–368.

Wade, M. J. 1982b. The evolution of interference competition by individual, family, and group selection. Proceedings of the National Academy of Science USA 79:3575–3578.

Wade, M. J. 1982c. Group selection: migration and the differentiation of small populations. Evolution 36:949–961.

Wade, M. J. 1984a. The changes in group-selected traits that occur when group selection is relaxed. Evolution 38:1039–1046.

Wade, M. J. 1984b. The influence of multiple inseminations and multiple foundresses on social evolution. Journal of Theoretical Biology 112:109–121.

Wade, M. J. 1984c. Variance effective population number: the effects of sex ratios and density on the mean and variance in offspring numbers in the flour beetle, *Tribolium castaneum*. Genetical Research 43:249–256.

Wade, M. J. 1985a. The effects of genotypic interactions on evolution in structured populations. In: Genetics: new frontiers, proceedings of the XV International Congress on Genetics, 283–290. Oxford University Press, Oxford, UK.

Wade, M. J. 1985b. Soft selection, hard selection, kin selection, and group selection. American Naturalist 125:61–73.

Wade, M. J. 1990. Genotype environment interaction for climate and competition in a natural population of *Tribolium castaneum*. Evolution 44:2004–2011.

Wade, M. J. 1992. Sewall Wright: gene interaction in the shifting balance theory. In: Oxford surveys of evolutionary biology, vol. 6, edited by J. Antonovics and D. Futuyma, 35–62. Oxford University Press, New York.

Wade, M. J. 1994. The biology of the willow leaf beetle, *Plagiodera versicolora* (Laicharting). In: Novel aspects of the biology of Chrysomelidae, edited by P. Joliviet and M. Cox, 541–547. Kluwer Academic, Dordrecht, The Netherlands.

Wade, M. J. 1995. Mean crowding and sexual selection in resource polygynous mating systems. Evolutionary Ecology 9:118–124.

Wade, M. J. 1996. Adaptation in subdivided populations: kin selection and interdemic selection. In: Evolutionary biology and adaptation, edited by M. R. Rose and G. Lauder, 381–405. Sinauer Associates, Sunderland, MA.

Wade, M. J. 1998. The evolutionary genetics of maternal effects. In: Maternal effects as adaptations, edited by T. A. Mousseau and C. Fox, 5–21. Oxford University Press, Oxford, UK.Wade, M. J. 2000. Epistasis: genetic constraint within populations and accelerant of divergence among them. In: Epistasis and the evolutionary process, edited by J. Wolf, E. Brodie III, and M. J. Wade, 213–231. Oxford University Press, Oxford, UK.

Wade, M. J. 2001. Epistasis, complex traits, and rates of evolution. Genetica 112:59–69.

Wade, M. J. 2002. A gene's eye view of epistasis, selection, and speciation. Journal of Evolutionary Biology 15:337–346.

Wade, M. J. 2013. Phase III of Wright's shifting balance process increases the genetic variation among demes for non-fitness traits. Evolution 67:1591–1597.

Wade, M. J. 2014. Paradox of Mother's Curse and the maternally provisioned offspring microbiome. In: The genetics and biology of sexual conflict, edited by W. Rice and S. Gavrilets, 73–81. Cold Spring Harbor Laboratory Press, Cold Spring Harbor, NY.

Wade, M. J., and R. W. Beeman. 1994. The population dynamics of maternal effect selfish genes. Genetics 138:1309–1314.

Wade, M. J., P. Bijma, E. D. Ellen, and W. Muir. 2010. Group selection and social evolution in domesticated animals. Evolutionary Applications 3:453–465.

Wade, M. J., and F. J. Breden. 1980. The evolution of cheating and selfish behavior. Behavioral Ecology and Sociobiology 7:167–172.

Wade, M. J., and F. J. Breden. 1981. The effect of inbreeding on the evolution of altruistic behavior by kin selection. Evolution 35:844–858.

Wade, M. J., and F. J. Breden. 1986. Life history of natural populations of the imported willow leaf beetle, *Plagiodera versicolora* (Coleoptera: Chrysomelidae). Annals of the Entomological Society of America 79:73–79.

Wade, M. J., and F. J. Breden. 1987. Kin selection in complex groups: mating structure, migration structure, and the evolution of social behaviors. In: Migration and social behavior, edited by Z. Halperin and D. Chepko-Sade, 273–283. University of Chicago Press, Chicago.

Wade, M. J., and C. J. Goodnight. 1991. Wright's shifting balance theory: an experimental study. Science 253:1015–1018.

Wade, M. J., and C. J. Goodnight. 1998. Genetics and adaptation in metapopulations: when nature does many small experiments. Evolution 52:1537–1553.

Wade, M. J., and J. R. Griesemer. 1998. Populational heritability: empirical studies of evolution in metapopulations. American Naturalist 151:135–147.

Wade, M. J., and D. E. McCauley. 1980. Group selection: the phenotypic and genotypic differentiation of small populations. Evolution 34:799–812.

Wade, M. J., and D. E. McCauley. 1984. Group selection: the interaction of local deme size and migration on the differentiation of small populations. Evolution 38:1047–1058.

Wade, M. J., and D. E. McCauley. 1988. The effects of extinction and colonization on the genetic differentiation of populations. Evolution 42:995–1005.

Wade, M. J., N. K. Priest, and T. Cruickshank. 2009. A theoretical overview of maternal genetic effects: evolutionary predictions and empirical tests using sequence data within and across mammalian taxa. In: Maternal effects in mammals, edited by D. Maestripieri and J. M. Mateo, chapter 3. University of Chicago Press, Chicago.

Wade, M. J., S. M. Shuster, and J. Demuth. 2003. Sexual selection favors female-biased sex ratios: the balance between the opposing forces of sex-ratio selection and sexual selection. American Naturalist 162:403–414.

Wade, M. J., S. M. Shuster, and L. Stevens. 1996. Inbreeding: its effect on response to selection for pupal weight and the heritable variation in fitness in the flour beetle, *Tribolium castaneum*. Evolution 50:723–733.

Ward, P. S. 1983. Genetic relatedness and colony organization in a species complex of ponerine ants. Behavioral Ecology and Sociobiology 12:301–307.

Werfel, J., and Y. Bar-Yam. 2004. The evolution of reproductive restraint through social communication. Proceedings of the National Academy of Sciences USA 101:11019–11024.

West, S. A., A. Gardner, and A. S. Griffin. 2006. Altruism. Current Biology 16:R482–483.

West-Eberhard, M. J. 1975. The evolution of social behavior by kin selection. Quarterly Review of Biology 50:1–33.

Whitlock, M. C. 1992. Non-equilibrium population structure in forked fungus beetles: extinction, colonization, and the genetic variance among populations. American Naturalist 139:952–970.

Whitlock, M. C., and D. E. McCauley. 1990. Some population genetic consequences of colony formation and extinction: genetic correlations within founding groups. Evolution 44:1717–1724.

Wiens, J. A. 1966. On group selection and Wynne-Edwards hypothesis. American Scientist 54:273–287.

Wild, G., A. Gardner, and S. A. West. 2009. Adaptation and the evolution of parasite virulence in a connected world. Nature 459:983–986.

Williams, G. C. 1966. Adaptation and natural selection. Princeton University Press, Princeton.

Williams, G. C. 1975. Sex and evolution. Princeton University Press, Princeton.

Williams, G. C. 1988. Retrospect on sex and kindred topics. In: The evolution of sex: an examination of current ideas, edited by R. E. Michod and B. R. Levin, 287–298. Sinauer Associates, Sunderland, MA.

Williams, G. C., and D. C. Williams. 1957. Natural selection of individually harmful social adaptations among sibs with special reference to social insects. Evolution 11:32–39.

Wilson, D. S. 1979. Structured demes and trait-group variation. American Naturalist 113:606–610.

Wilson, D. S. 1980. The natural selection of populations and communities. Benjamin/Cummings, Menlo Park, CA.

Wilson, D. S. 1983a. The group selection controversy: history and current status. Annual Review of Ecology and Systematics 14:159–187.

Wilson, D. S. 1983b. The effect of population structure on the evolution of mutualism: a field test involving burying beetles and their phoretic mites. American Naturalist 121:851–870.

Wilson, D. S., and E. Sober. 1994. Reintroducing group selection to the human behavioral sciences. Behavioral and Brain Sciences 17:585–608.

Wilson, E. O. 1971. The insect societies. Harvard University Press, Cambridge, MA.

Wilson, E. O. 1975. Sociobiology: the new synthesis. Harvard University Press, Cambridge, MA.

Wilson, E. O., and B. Hölldobler. 2005. Eusociality: origin and consequence. Proceedings of the National Academy of Sciences USA 102:13367–13371.

Wimsatt, W. C. 1980. Reductionistic research strategies and their biases in the units of selection controversy. In: Scientific discovery: case studies, edited by P. Kitcher, 213–259. Springer, Dordrecht, The Netherlands.

Wimsatt, W. C. 2007. Re-engineering philosophy for limited beings: piecewise approximations to reality. Harvard University Press, Cambridge, MA.

Winther, R., C. Dimond, and M. J. Wade. 2013. The Fisher-Wright debate and the units of selection controversy: differing aims, complementary analyses, and unified evolutionary theory. Philosophy of Biology 28:957–979.

Wolf, J. B., E. D. Brodie III, J. M. Cheverud, A. J. Moore, and M. J. Wade. 1998. Evolutionary consequences of indirect genetic effects. Trends in Ecology and Evolution 13:64–69.

Wolf, J. B., E. Brodie III, and M. J. Wade, eds. 2000. Epistasis and the evolutionary process. Oxford University Press, Oxford, UK.

Wolf, J. B., and J. M. Cheverud. 2012. Detecting maternal-effect loci by statistical cross-fostering. Genetics 191:261–277.

Wolf, J. B., T. T. Vaughn, L. S. Pletscher, and J. M. Cheverud. 2002. Contribution of maternal effect QTL to genetic architecture of early growth in mice. Heredity 89:300–310.

Wolf, J. B., and M. J. Wade. 2001. On the assignment of fitness to parents and offspring: whose fitness is it and when does it matter? Journal of Evolutionary Biology 14:347–356.

Wolf, J. B., and M. J. Wade. 2009. What are maternal effects (and what are they not)? Philosophical Transactions of the Royal Society B 364:1107–1115.

Wright, S. 1931. Evolution in Mendelian populations. Genetics 16:97–159.

Wright, S. 1932. The roles of mutation, inbreeding, crossbreeding, and selection in evolution. Proceedings of Sixth International Congress on Genetics 1:356–366.

Wright, S. 1938. Size of population and breeding structure in relation to evolution. Science 87:430–431.

Wright, S. 1943. Isolation by distance. Genetics 28:114–138.

Wright, S. 1945. Tempo and mode in evolution: a critical review. Ecology 26:415–419.

Wright, S. 1959. Physiological genetics, ecology of populations, and natural selection. Perspectives in Biology and Medicine 3:107–151.

Wright, S. 1969. Evolution and the genetics of populations. Vol. 2: The theory of gene frequencies. University of Chicago Press, Chicago.

Wright, S. 1977. Evolution and the genetics of populations. Vol. 3: Experimental results and evolutionary deductions. University of Chicago Press, Chicago.

Wright, S. 1978. Evolution and the genetics of populations. Vol. 4: Variability within and among natural populations. University of Chicago Press, Chicago.

Wright, S., and W. E. Kerr. 1954. Experimental studies of the distribution of gene frequencies in very small populations of *Drosophila melanogaster*, II: Bar. Evolution 8:225–240.

Wynne-Edwards, V. C. 1962. Animal dispersion in relation to social behaviour. Oliver and Boyd, Edinburgh, UK.

Wynne-Edwards, V. C. 1964. Population control in animals. Scientific American 211:68–74.

Wynne-Edwards, V. C. 1965. Self-regulating systems in populations of animals. Science 147:1543–1548.

Index

Page numbers in italics refer to illustrations.

dominance, 219

Drosophila (fruit flies): *D. melanogaster*, 84, 120, 228; electrophoresis protocols for, 94; random drift studies on, 95; visible mutations of, 84; and Wade's dissertation, 56

ecosystems, "islands" view of, 48

electrophoresis, 49, 94

Emerson, A. E., 72, 163

ENCODE (Encyclopedia of DNA Elements), 7

environmental variation, 41–43, *42*; effects of, on heritability, 115–16; and efficacy of group selection, 135–36; and fitness levels, 158; scale of, 132–35, *133*

epistasis. *See* gene interactions

eugenics, 52

evolution: gene's eye view of, 23, 76, *77*, 79; requisites of, 3; by single genes, 4; and Wright's Shifting Balance Theory, 190–91

Evolution, 100

Evolutionary Ecology (Pianka), 115

evolutionary genetics field, 45–46

Evolution in Changing Environments (Levins), 48

Ewens, W., 192

external environment, genetic adaption to, 7

"Extinction" (Levins), 48

extinctions: and competitive ability, 122–23; differential extinction, 183, 191; and fitness variation, 82; and g^2 (group/population heritability), 97; and heritable variation, 99; in island biogeography, 50; and phenotypic variation, 80, 81; and rate of population growth (r), 54, 123; and Wade's dissertation, 56, 59, 60

family selection: and breeding practices, 23, 41, 75–76; definition of, 29; and environmental variation, 41–43; and gene interactions, 43; and genetic variation, 36–37; and individual selection, 32, *33*; and kin selection theory, 77, 78–79; and magnitude of evolutionary change, 37; and rate of population growth (r), 77; with simple genetics, 35–37; and social interactions, 43–44; and survival/reproduction variations, 30–31, *31*

fecundity, 30–31, *31*, 63, 64–65

Feldman, M., 6, 100

fertilization and cryptic female choice, 15

Fisher, R. A.: and debates on group selection, 27; and gene interactions, 85; and G × G, 43; and Hamilton's paper on kin selection, 138; on inbreeding vs. selection, 150; on natural populations, 28; on sexual selection, 75, 134, 153–54

Fisher's Fundamental Theorem, 49, 52, 84, 189–90, 191–93

fitness: and altruism, 153–54, 156, 157, 158; and antagonistic pleiotropy, 114; and conversion effect, 12; declines in, as by-product of selection, 114; fitness variation, 82–84; and g^2 (group/population heritability), 11; and gametic selection (also known as meiotic drive), 15; genetic variation for individual fitness $V(G_I)$, 222–26, *223, 224, 225*; and group selection for individual traits, 28; and inbreeding, 114; inclusive-fitness theory, 162; indirect effects, 156, 157; individual as related to population, 85–86; and kin selection theory, 51–52; marginal gene fitness, 76–77; relative, 88, 196; and selection among groups, 16; and sex ratio, 24; and social evolution, 156–60; and sterile castes problem, 23; and strength of selection, 88; and variations in survival and reproduction, 30–31; and Wade's dissertation, 58–59

flour beetles. See *Tribolium*

founder number: ecological vs. genetic effects on, 125–32, *127, 128, 129, 130*; and growth rate, 116; and population size, 118

fraction of genetic variation among populations. See F_{st}

Franklin, I., 48

freedom of speech example, 18

"free rider" problem, 18

frequency-dependent selection, 49

fruit fly studies, 51. See also *Drosophila*

F_{st} (fraction of genetic variation among demes): as fixation index, 92–93; and g^2 (group/population heritability), 103; and genetic relatedness (r), 50–51; and "haystack model" of group selection, 94; and island-model migration, 101;

and breeding practices, 9, 41–44; and cannibalism, 181; and competition, 40, 110, 114–15; and Darwinian natural selection, 17; debates on, 21–22, 25–29, 48–49, 136; and density regulation, 37–39, 38; by differential dispersion, 19, 183, 184, 191, 194, 199; by differential extinction, 183, 191; and discontinuities, 50; ecology of, 113–17; and effects of adaptation on individuals/groups, 17–19; efficacy of, 99, 135–36; and fitness variation, 83–84; formation of groups, 16; and gametic selection, 6–7; "haystack model" of, 93–94; and heritability, 10 (see also g^2); and heterogeneity, 121; and individual selection, 134; for individual traits, 27–29; and island-model migration, 101; and kin selection, 52, 76–78, 137–41, 140, 149–51, 152; and life-history theory, 49; and loss of gains, 113–14; and metapopulations, 48; and migration, 17, 61–62, 93–94, 103–4; in Muir's poultry experiment, 9; and multilevel selection theory, 8; and population mean fitness, 113; and r- and K-populations, 120; and random drift, 219; and rate of population growth, 54–56; and reductionist biases, 79–80; relaxing, 117–21, 135–36; self-facilitating aspects of, 14; as selfish herd, 27; and sex ratio, 23–24; and social interactions, 43–44; standard theory on, 4; and sterile castes, 22–23; and tenure application, 163; and *Tribolium* findings, 107; and $V(Z)$, 80; and variation, 80; and Wade's dissertation, 54–57, 58–60; Wade's interest in, 47; and willow leaf beetle studies, 180–82; and Wright's Shifting Balance Theory, 19

h^2 (individual heritability): and conversion effect, 12; definition of, 10–11; and environmental variation, 115–16; experiments, 99; and g^2 (group/population heritability), 11, 59, 85, 86–87; and gene interactions, 12, 85, 114; and genetic relatedness (r), 182; origins of, in interacting genetic systems, 12; term, 10; and willow leaf beetle studies, 182

Hager, R., 142

Haldane, J. B. S., 14–15, 78, 138, 143–44, 146

Hamilton, W. D.: "The genetical evolution of social behavior," 16; haplodiploid model of, 143; and inclusive-fitness theory, 156; on kin selection and altruism, 137–38

Hamilton's Rule, 138–39; and cheaters, 146, 147; and evolution of altruism, 141–42; within groups, 148–49; and inclusive-fitness theory, 159; and multilevel selection theory, 159

haplodiploidy, 143

hard selection, 32, 37–39

Hartl, D., 100

"haystack model" of group selection, 93–94

Heisler, I. Lorraine, 132, 134

hens, 8–9, 158–59

heredity: and gametic selection, 14; and heritability as multilevel process, 11; as requisite of evolution, 3; and selection, 179; single genes as conveyors of, 4

heritability, 91–111; expressed as a covariance, 88; genetic differentiation vs., 96–97; in individuals (see h^2); and multilevel selection theory, 10–13; in populations or groups (see g^2); "realized" (see $g^2(t)$)

heritable variation, 55, 59, 125

heterogeneity, 121

Hill-Robertson effect, 14

homozygosity, 99, 219

"Host plants as islands in evolutionary and contemporary time" (Janzen), 48

Huang, W. S., 228

Huffaker, C. B., 46

human genome, regulatory regions of, 7

inbreeding: and altruism, 151–53, 154–55, 155; consequences of, 32, 34, 36; ecological reasons for, 39; Fisher on, 150; and fitness declines, 114; and genetic variation, 36; and inclusive-fitness theory, 161–62; and multilevel selection theory, 162; and purging hypothesis for SBT finding, 219–21, 221; and rate of social evolution, 149–51; and run-away sexual selection, 154; in *Tribolium* studies, 99–100; and varying N without m experiment, 97; and within-family selection, 32, 34, 40–41 inclusive-fitness theory: and altruism, 77; and Central Theorem of Sociobiology, 77; and Hamilton's Rule, 159; and inbreeding, 161–62; and incorrect predictions, 162; and kin selection theory, 51; and methodological individualism, 160; and multilevel selection theory, 21–22, 159, 160–61; and social evolution, 156–57; and sterile castes problem, 23

Indiana University, 1, 226

indirect genetic effects (IGEs): and competitive interactions, 39–41; Griffing's investigation of, 85; and heritability, 10; and maternal genetic effects, 230; and polyandry, 40; and variations in survival and reproduction, 30

individual heritability. See h^2

individual selection: and altruism, 142; and among-family selection, 37; and breeding practices, 9, 23; and cannibalism, 181; and cattle, 75; and competition, 110; and Darwinian natural selection, 23; and effects of adaptation on individuals/groups, 17–19; and efficacy of group selection, 99; and environmental variation, 41–43; and family selection, 32, 33; and fitness variation, 83–84; and future efficacy, 211; and gene interactions, 43; and group selection, 134; and "haystack model" of group selection, 93–94; and heritability, 10, 215 (see also h^2); and inclusive-fitness theory, 77; and kin selection, 16, 52; and magnitude of evolutionary change, 37; and Muir's poultry experiment, 9; privileged position of, 74; and r- and K-populations, 120; and rate of population growth, 55–56; self-limiting aspects of, 14; and sex ratio, 23–24; and social interactions, 43–44; standard theory on, 4; unwanted by-products of, 114; and $V_{ave}(z)$ [metapopulation average of $V(z)$], 80; and within-family selection, 37; and Wright's Shifting Balance Theory, 19, 210–11

individual traits, group selection for, 27–29

Inger, R., 53, 67

Insect Societies, The (Wilson), 48

interactions: and additive genetic variation, 12–13; among levels-of-selection, 13–14, 43; in card game analogy, 5–6, 8; and conversion effect, 12; and h^2 (individual heritability), 12; interaction effects, 85; and multilevel selection theory, 7–8, 13–14; and selection below the level of the individual, 14–15; standard theory on, 4

internal environment, genetic adaptation to, 7

interpopulation theory of Wright, 183. *See also* group selection

Introduction to Population Genetics Theory, An (Crow and Kimura), 48

island biogeographic theory, 45–46, 50

island-model migration, 79–80, 81, 101–4, 102, 105, 106

isolation by distance, 101

"Is the gene the unit of selection?" (Franklin and Lewontin), 48

Janzen, D., 47, 48, 91

Kalisz, Sue, 132

Kauffman, S., 47

Kelly, Carol, 170

Kiester, Ross, 72

Kimura, M., 4, 48

kin selection, 50–53; as alternative to group selection, 76–78; and altruism, 143, 151–53; and evolution of sociality, 16; and family selection, 77, 78–79; influence of, 163–64; and mutations, 146–47; and rate of social evolution, 149–51; as related to group selection, 137–41, 140, 149–51, 152

Kover, Paula, 226

Kreitman, M., 213

K-selected species, 120

Lack, D., 25

Lande, Russ, 87, 120, 135, 163

Leibold, M., 213

levels-of-selection, 13–14, 41–44, 42, 55, 79

Levins, R.: and academic tradition of University of Chicago, 72; on dynamic metapopulations, 54; *Evolution in Changing Environments*, 48; "Extinction," 48; on group selection, 49; on metapopulations, 6, 16; and population biology, 47; status of, in field, 91; and Wade's dissertation, 58–59

Lewontin, R. C.: and academic tradition of University of Chicago, 72; departure of, 71; influence of, 91, 163; "Is the gene the unit of selection?," 48; and population genetics, 47; and reductionist biases against group selection, 79; on selection below the individual, 6; "Units of selection," 48

life-history theory, 49

Lillie, F. R., 72

livestock breeders, 8

Lloyd, M. B., 47, 67, 91

local fitness optima, 186

local populations, 27, 39

Lopez-Fanjul, C., 195

Lucas, Ora Lee, 123, 184, 226, 227

m (rate of migration): and genetic differentiation vs. heritability, 96–97; and Smith's "haystack model" of group selection, 93; varying both m and N, 104–7; varying m while keeping N constant, 101–4, 102; varying N without m, 97–100

MacArthur, R. H., 48, 115

MacKay, T., 229

macroevolutionary processes, 7–8

marginal gene fitness, 76–77

mass selection, 186. See also individual selection

maternal genetic effects, 230

mating systems: and altruism, 151–53, 152; and cannibalism, 139–41, 140, 149–50; coevolution of sociality and, 153–55; and direct vs. indirect genetic effects, 40; and genetic relatedness (r), 175–78; and genetic variation, 36–37; and rate of social evolution, 149–50, 151–53. See also inbreeding

Mayr, E., 52

McCauley, David E.: on academic culture at University of Chicago, 47; allergies of, 165; collaboration with Wade, 80, 132–33; on efficacy of group selection, 199; on extinction rates, 97; and founder number studies, 125; on F_{st} (fraction of genetic variation among demes), 50, 94; and "Persistent" metapopulations, 123; and reactions to concept of group

heritability, 100; and "sacrifice" design, 172; and soldier beetle studies, 106–7; on sperm carry-over effects, 96; on variation in population growth rates, 65; and varying N without m experiment, 98; and Wade's doctoral research, 68; and willow leaf beetle studies, 171, 175, 176

McElroy, Pat, 164

MEDEA, 144

Mendelian genetics, 77

Mendel's Law of Segregation, 14

Mertz, D. B., 48, 53, 56, 67, 185

metapopulations: definition of, 6; development of concept, 48

methodological individualism, 160

microevolutionary processes, 7–8

migration: and competition, 188; effective migration rate, 207–9; and fitness variation, 82; and group heritability (g^2), 60; group members exchanged through, 17; and group selection, 28, 61–62, 93–94, 103–4; and individual selection, 74; island-model migration, 79–80, 81, 101–4, 102, 105, 106; and phenotypic variation, 80–82; rate of, 93, 96–100, 101–7, 102; stepping-stone migration, 101, 106; and Tribolium model, 207–10, 209; and varying m while keeping N constant, 101–4; and Wade's dissertation, 60–61; and Wright's Shifting Balance Theory, 187–88, 191, 194–96, 197–98, 199, 199–200, 202–5, 207–10, 209, 215–16

milkweed beetles (Tetraopes tetraophthalmus), 132

mitochondrial defect of cytoplasmic male sterility (CMS), 15–16

monogamy, 36–37

mortality, genetic and ecological causes of, 37–39

mosquitoes, 148

Mueller, L. D., 120

Muir, Bill, 9

multifamily groups, 142–43

multilevel selection theory, 7–8; and artificial group selection, 8; and cannibalism, 179–82; and cheaters, 160; comprehensive, 6–7; and debates on group selection, 21–22; development of, 6; and effect of social environment on fitness,

multilevel selection theory (*continued*)
157–60; and family selection, 29; and
gametic selection, 6–7; and genetic relatedness (*r*), 159; and Hamilton's Rule,
159; and heredity, 9–13; and inbreeding,
162; and inclusive-fitness theory, 21–22,
159, 160–61; and interactions, 7–8, 13–14;
and kin selection theory, 78; and literature on artificial selection, 76; and predictions of variations within/between
populations, 29–30; and sex ratio, 24
mutation, 74, 145–47

Nathanson, M. R., 54, 56
National Evolutionary Synthesis Center, 1
National Institutes of Health (NIH), 3, 100,
131, 213, 226
National Science Foundation, 1, 100
natural selection: and conversion effect, 12;
and environmental fit of organism, 22;
and Fisher's Fundamental Theorem,
191–92; and group selection, 17, 27–28;
and heritable variation in fitness, 84;
and mutation, 145; and reproduction
limitation, 25–26; selection and heritability processes of, 87; and selective
gains, 113; and selfishness, 26; and sex
ratio, 23–24; and Wright's Shifting Balance Theory, 191
"Natural selection of individually harmful
social adaptations among sibs with
reference to social insects" (Williams),
78–79
Neutral Theory of Evolution, 4
nine-banded armadillos, 52
Nonacs, P., 142
"Non-Darwinian Evolution" (King and
Jukes), 4

O'Donnell, R., 171, 175
one-gene-at-a-time model, 4, 19, 136
Opportunities for Promoting Understanding
through Synthesis (OPUS) Program of
the National Science Foundation, 1
outbreaks in population *N*, 118–19

Park, Thomas: and academic tradition of
University of Chicago, 72; and cannibalism, 64; on competition, 46, 62, 114;

control population protocol, 117; field
ecology course of, 53; on founder number, 116; influence of, 163; and kin selection, 139; and letters of recommendation, 70; and MacArthur, 115; population
ecology course of, 47–48; retirement
of, 69; support of Wade, 70, 70; and
Tribolium studies, 62, 69; and Wade's
beetle research, 117; Wade's competition
with, 163–64; and Wade's dissertation,
3, 53, 54, 55–57, 60, 61, 67, 67; and Wade's
teaching career, 91; and Wright's Shifting Balance Theory, 188, 193
parsimony principle, 17, 48, 55, 74, 78
Peabody Museum of Yale University, 2
Pearl, Raymond, 164
Perkins, Marlin, 2
Pfister, K., 213
phenotypic variation among populations,
80–82
phosphoglucomutase (*pgm*), 175
Pianka, E. R., 115
Plagiodera versicolora (willow leaf beetles),
116
plants, leaf area of, 132–34, *133*
Plocke, Donald J., 3
poker and war (card games) analogy, 5–6, 8,
42, 85
pollen competition, 15
polyandry, *35*, 36–37, *40*
polymorphism, 49, 50, 94, 179
population growth, rate of (*r*): application
of, in research fields, 49–50; and competitive ability, 121–23, *122*; ecological
effect on, 125–32, *127*; and extinctions,
123; and family selection, 77; and individual selection, 55–56; and population
mean fitness, 107–10; and relaxation of
group selection, 121; and *Tribolium* experiments, 54–55, 107–10; and Wade's
dissertation, 60
population heritability. See *g²*
"Post-doc Breakfast Club," 137
postman problem, 188
poultry farming, 8–9, 158–59
predator-prey theory, 46–47

*Quantitative Genetic Studies of Behavioral Evo-
lution* (Boake), 135

sociality and social evolution (*continued*)
143–47, 160–61; coevolution of mating
system and, 153–55; evolution of, 16; and
frequency distribution of social genes,
161; and genetic relatedness (*r*), 50–51;
and Hamilton's Rule, 137–38, 141–42, 146,
147, 148–49, 159; and inbreeding, 149–51,
161–62; and inclusive-fitness framework,
156–57; and incorrect predictions, 162;
and kin selection, 16, 51, 137–42, 143, 146–
47, 149–51; mating and associating with
kin, 151–53; and multilevel selection
framework, 157–60; and rate of social
evolution, 149–50; and selection beyond
two levels, 147–49; and variation at different spatial scales, 178–79. *See also*
altruism
social learning, 188
sociobiology, 52, 77–78
sociogenomics, 161
soft selection (within-family selection), 32,
34, 34–37, 39–43
soldier beetles (*Chaliognathus pennsylvanicus*), 106–7, 132
sperm carry-over effects, 96
Spiess, E., 72
stepping-stone migration, 106
sterile castes, 17, 22–23, 24
Stevens, L., 176, 185
strength of selection [$S(t)$], 87, 157, 201
Swift, Hewson, 72

Taylor, C. E., 120
Tetraopes tetraophthalmus (milkweed beetles),
132
Theory of Island Biogeography, The (MacArthur
and Wilson), 48
Tinley Creek Forest Preserve, 7, 165
Tonsor, Steve, 132
trait function(s), 22
traveling salesman problem, 188
Tribolium (flour beetles): and cannibalism,
50, 109, *109*, 110, 138–41, *140*, 182; c-ARK
population, 123, 126; competition
studies using, 46, 121–23; c-SM stock,
115, 123; and ecology/genetics interaction, 131; and ecology/genetics of
population growth, 129–30; and efficacy
of group selection, 135; electrophore-

sis protocols for, 94; and environmental variation, 116; and family selection,
29; and founder number studies, 125–28;
and genetic relatedness (*r*), 182; kin
selection study, 138–41, *140*; Nathanson's study on, 54; and N_e (effective size),
92–96; and NIH funding, 226; Park's
studies of, 46, 54, 62, 69; polymorphism estimates for, 94; and population
growth rate, 121–23; reduced competitive ability in, 114; and relaxing selection, 117–21; sensitivity to F_{ST}, 106; and
sperm carry-over effects, 96; suitability
of, for research purposes, 54–55, 69, 100;
T. castaneum, 55, 69, 123–25; *T. confusum*,
55, 69, 122, 124; traits of, 62, 63; and
variation for fitness, 123–25; and Wade's
allergies, 165; and Wade's doctoral research, 54–68, *57*, 79; and willow leaf
beetle studies, 182; and Wright's Shifting Balance Theory (see *Tribolium* model
of Wright's Shifting Balance Theory)
Tribolium experiments, 96–111; and cannibalism, 109, *109*, 110; and population mean
fitness, 107–10; and Slatkin's theoretical findings, 107; varying both *m* and *N*,
104–7; varying *m* while keeping *N* constant, 101–4, *102*; varying *N* without *m*,
97–100
Tribolium model of Wright's Shifting Balance
Theory, 193–211; control for Phase III
group selection, 197–99, *199*; and effect
of migration on mean fitness, 215–16,
217, 217–19, *218*; experimental design,
194–97, *197*; and interdemic selection,
219; and purging hypothesis for SBT
finding, 219–21, *221*; reception of findings, 212; response to Phase III selection,
200–202, *207*; results of, 202–11, *203*, *205*
Trivers, R., 138

"Units of Selection, The" (Lewontin), 6, 48
University of Chicago: academic tradition at,
72–73; career options of Wade at, 71; and
chair appointment, 213; Department of
Ecology and Evolution (E & E), 213–14;
Department service at, 92; education of
Wade at, 3, 45, 47; and graduate program
rankings, 214; position with, 72–73, *73*;

Wright's Shifting Balance Theory (*continued*)
ment of, 27; and Fisher's Fundamental
Theorem, 191–93; fitness as focal trait of,
189; and F_{st} (fraction of genetic variation
among demes), 93; and gene-interaction
emphasis of Wright, 186–90, 191, 192,
195; and interdemic selection mechanism, 183–84, 187, 195, 199–200, 201–2,
206–7, *207*, 210–12, 219 (*see also* group

selection); three phases of, 190–91; and
Tribolium studies (see *Tribolium* model
of Wright's Shifting Balance Theory);
Wade's review of, 193, *194*

Wu, C.-I., 213

Wynne-Edwards, V. C., 25, 47, 74–75

Yale University, 2

Yerkes, R. M., 47